NATIONAL AERONAUTICS AND SPACE ADMINISTRATION

Technical Report No. 32-1023

Surveyor I Mission Report

Part I. Mission Description and Performance

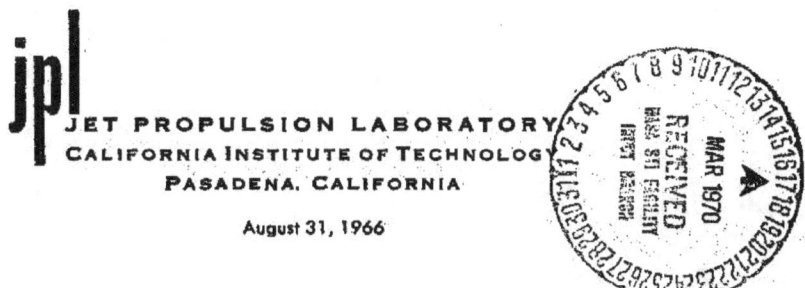

JET PROPULSION LABORATORY
CALIFORNIA INSTITUTE OF TECHNOLOGY
PASADENA, CALIFORNIA

August 31, 1966

ISBN: 9798861270632
Imprint: Independently published

JPL/NASA 1966
Prepared Linder Contract No. NAS7-100
National Aeronautics & Space Administration

PREFACE

This three-part document constitutes the Project Mission Report on *Surveyor I*, the first in a series of unmanned lunar soft-landing missions.

Part I of this report consists of a technical description and an evaluation of engineering results of the systems utilized in the *Surveyor I* mission. Part I was prepared from the contributions of a number of individuals who were cognizant over particular systems. Much of the content of the report was drawn from individual systems reports and other published documents. Analysis of the data received from the *Surveyor I* mission is continuing and some improvement in accuracy is expected in the specific performance values presented here.

Part II of this report comprises the scientific analysis and interpretation of the video and other data gained from the mission. Part III is a compilation of the data used for the scientific analysis, including selected photographs.

CONTENTS

CONTENTS (Cont'd)

TABLES

TABLES (Cont'd)

FIGURES

FIGURES (Cont'd)

FIGURES (Cont'd)

FIGURES (Cont'd)

ABSTRACT

Surveyor I, the first of a series of unmanned, soft-landing missions, was launched from Cape Kennedy, Florida, on May 30, 1966, and achieved a perfect soft-landing on the moon, June 2, 1966. All Project and Flight Objectives for this mission were satisfied. The spacecraft (*Surveyor I*) continued to operate successfully and provide telemetry data until two days after sunset of the first lunar day, when telemetry transmission was purposely discontinued for the lunar night. Spacecraft operation was resumed during the second lunar day and continued performing until sunset, July 14, 1966, when the operational phase of the *Surveyor I* mission was terminated. Over 100,000 ground commands were transmitted to the spacecraft during the course of the mission, and over 11,000 pictures were returned and recorded. A large quantity of data was also received related to lunar bearing strength, radar reflectivity, and surface temperature of the moon. A technical description of the mission and an evaluation of the engineering results are presented herein.

I. INTRODUCTION

Surveyor I was launched on Mission A from Cape Kennedy, Florida, at 14:41:00.990 GMT on May 30, 1966, and soft-landed on the moon at 06:17:37 GMT on June 2, 1966. All established goals were attained on this mission, the first in a series of unmanned spacecraft flights designed to soft-land and function on the lunar surface. In addition to meeting established objectives, *Surveyor I* demonstrated a capability to operate successfully in the lunar environment over a period of time greatly in excess of that required by the mission objectives—including survival through the long lunar night and resumption of operations during the following period of daylight.

At the termination of *Surveyor I* operations at the close of the second lunar day, on July 14, 1966, over 100,000 commands had been sent to the spacecraft and over 11,000 pictures had been returned.

A. Surveyor *Project Objectives*

Surveyor is one of two unmanned lunar exploration projects currently being conducted by the National Aeronautics and Space Administration. The other, *Lunar Orbiter*, is planned to provide medium- and high-resolution photographs over broad areas to aid in site selection for the *Surveyor* and *Apollo* landing programs.

The overall objectives of the *Surveyor* Project are:

1. To accomplish successful soft landings on the moon as demonstrated by operations of the spacecraft subsequent to landing.

2. To provide basic data in support of *Apollo*.

3. To perform operations on the lunar surface which will contribute new scientific knowledge about the moon and provide further information in support of *Apollo*.

The first seven flights, identified as Missions A through G, are currently authorized in the *Surveyor* Project. The first four of these missions were planned to satisfy Project Objective 1 above. The last three were planned to meet Project Objective 2 above. Planning and engineering are underway for additional missions, which are intended to meet Objective 3 above. However, no flight hardware is currently authorized for this configuration.

The highly successful *Surveyor I* mission has satisfied Objectives 1 and 2 above, and has contributed significantly to the attainment of Objective 3. The *Surveyor I* landing site appears to be suitable for *Apollo*, which will utilize final descent and landing system technology similar to that of *Surveyor*.

B. Project Description

The *Surveyor* Project is managed by the Jet Propulsion Laboratory for the NASA Office of Space Science and Applications. The Project is supported by four major administrative and functional elements or systems: Launch Vehicle System, Spacecraft System, Tracking and Data Acquisition System (T&DA), and Mission Operations System (MOS). In addition to overall project management, JPL has been assigned the management responsibility for the Spacecraft, Tracking and Data Acquisition, and Mission Operations Systems. NASA/Lewis Research Center (LeRC) has been assigned responsibility for the *Atlas/Centaur* launch vehicle system.

1. Launch Vehicle System

Atlas/Centaur launch vehicle development began as an Advanced Research Projects Agency program for synchronous-orbit missions. In 1958, General Dynamics/Convair was given the contract to modify the *Atlas* first stage and develop the *Centaur* upper stage, and Pratt & Whitney was given the contract to develop the high-impulse LH_2/LO_2 engines for the *Centaur* stage.

The Kennedy Space Center, Unmanned Launch Operations branch, working with LeRC, is assigned the *Centaur* launch operations responsibility. The *Centaur* vehicle utilizes Launch Complex 36, which consists of two launch pads (A and B) connected to a common blockhouse. The blockhouse has separate control consoles for each of the pads. Pad 36A was utilized for *Surveyor* Mission A.

Seven R&D flight tests were performed in the four years preceding the launch of *Atlas/Centaur* AC-10 for Mission A, which marked the first operational flight of the vehicle. The successful flight of AC-6 on August 11, 1965, demonstrated the operational capability for "direct ascent" missions. Direct ascent was the mode utilized for Mission A and will also be used for Missions B and D. One additional *Centaur* dual-burn development flight is scheduled to demonstrate capability to launch via "parking orbit ascent" trajectories. *Centaur* dual-burn flights are scheduled for Missions C, E, F, and G.

2. Spacecraft System

The *Surveyor* spacecraft weight of about 2200 lb and overall dimensions were established in accordance with the *Atlas/Centaur* vehicle capabilities. Three major features, first demonstrated on the *Ranger* missions, were incorporated in the *Surveyor* spacecraft system: fully attitude-stabilized spacecraft, earth-directed high-gain antenna, and the midcourse maneuver. Demonstration of TV communication at lunar distances is another *Ranger* achievement which has been of value to *Surveyor* and the other lunar programs. In addition, the *Surveyor* spacecraft utilizes several new features associated with the complex terminal phase of flight and soft-landing: throttleable vernier rockets with solid-propellant main retromotor; extremely sensitive velocity- and altitude-sensing radars, and an automatic closed-loop guidance and control system. The demonstration of these devices on *Surveyor* missions is a direct benefit to the *Apollo* program, which will employ similar techniques. Design, fabrication, and test operations of the *Surveyor* spacecraft are performed by Hughes Aircraft Company under the technical direction of JPL.

3. Tracking and Data Acquisition System

The T&DA system provides the tracking and communications link between the spacecraft and the Mission Operations System. For *Surveyor* missions, the T&DA system uses the facilities of: (1) the Air Force Eastern Test Range for tracking and telemetry of the spacecraft and vehicle during the launch phase, (2) the Deep Space Network for precision tracking, communications, data transmission and processing, and computing, and (3) the Manned Space Flight Network and the World-Wide Communications Network (NASCOM), both of which are operated by Goddard Space Flight Center.

The critical flight maneuvers and nearly all the picture-taking operations on Mission A were commanded and recorded by the Deep Space Station (DSS 11) at Goldstone, California, during its view periods. A few picture sequences were obtained by DSS 42, near Canberra, Australia, and the Johannesburg station (DSS 51), which served as prime stations for tracking and monitoring of engineering telemetry. The DSS stations near Madrid, Spain (DSS 61), at Ascension (DSS 72), and at Goldstone (DSS 14 with its large 210-ft antenna and DSS 12) were configured for monitoring and backup operation during the mission.

4. Mission Operations System

The Mission Operations System essentially controls the spacecraft from launch through termination of the mission. In carrying out this function, the MOS constantly evaluates the spacecraft performance and prepares and issues appropriate commands. The MOS is supported in its activities by the T&DA system as well as special hardware provided exclusively for the *Surveyor* Project and referred to as mission-dependent equipment. Included in this category are the Command and Data Handling Consoles installed in the DSS's, the Television Ground Data Handling System, and other special display equipment.

A JPL scan conversion system made it possible to display Mission A pictures in real-time on conventional TV monitors at the Laboratory as they were being relayed from the *Surveyor* on the moon via microwave from Goldstone. During the night of June 1 and the morning hours of June 2, when *Surveyor I* took its first 144 pictures, commercial television networks further relayed the "live" lunar program throughout the nation. The *Early Bird* satellite carried the pictures even farther—to Europe.

C. Mission A Objectives

All Mission A objectives were satisfied. The specific objectives of each *Surveyor* Mission are denoted as "flight objectives." For Mission A the flight objectives were specified in three categories: primary, secondary, and tertiary.

1. Primary Flight Objectives

    a. Demonstrate the capability of the *Surveyor* spacecraft to perform successful midcourse and terminal maneuvers and a soft-landing on the moon.

    b. Demonstrate the capability of the *Atlas/Centaur* vehicle to successfully inject the *Surveyor* spacecraft on a lunar intercept trajectory.

    c. Demonstrate the capability of the *Surveyor* Communications System and the Deep Space Network to maintain communications with the spacecraft during its flight and after the soft landing.

2. Secondary Flight Objectives

    a. Obtain in-flight engineering data on all spacecraft subsystems used in cruise flight.

    b. Obtain in-flight engineering data on all spacecraft subsystems used during the midcourse maneuver, terminal maneuver, and main retro phase.

    c. Obtain in-flight engineering data on the performance of the closed-loop terminal descent guidance and control system, consisting of the velocity and altitude radars, on-board analog computer, autopilot, and vernier engines.

    d. Obtain engineering data on the performance of spacecraft subsystems used on the lunar surface.

3. Tertiary Flight Objectives

    a. Obtain post-landing TV pictures of a spacecraft footpad and the immediately surrounding lunar surface material.

    b. Obtain post-landing TV pictures of the lunar topography.

    c. Obtain data on the radar reflectivity of the lunar surface.

    d. Obtain data on the bearing strength of the lunar surface.

    e. Obtain spacecraft temperature data on the lunar surface for use in the analysis of lunar surface temperatures.

Prior to Mission A launch, a launch-hold criterion was established in that the capability must exist for all project systems to meet all objectives (primary, secondary, and tertiary) before the launch would be permitted.

D. Mission A Summary

Surveyor I lifted off Pad 36A at Cape Kennedy within one second of its planned launch time at 14:41:01 GMT on May 30, 1966. The perfect countdown was followed by a boost into space by the *Atlas/Centaur* AC-10 vehicle that would have put the spacecraft on the moon within 250 miles of the aiming point. All launch vehicle and spacecraft events through Canopus acquisition occurred satisfactorily and at near-nominal times with the exception of the deployment of one of *Surveyor's* two low-gain antennas, Omniantenna A.

Two-way communications lock was achieved by Deep Space Station 51 (DSS-51) in Johannesburg, South Africa, at liftoff plus 28 min, transferring control of the mission from the Eastern Test Range to the Space Flight Operations Facility (SFOF) at JPL, in Pasadena, California. Commands were transmitted to *Surveyor* to deploy Omniantenna A, but telemetry indicated no change.

Although the minor antenna failure was not expected to seriously jeopardize the mission from either the standpoint of communications or because of the slight change in spacecraft center of gravity, the exact position of the antenna remained undetermined. This was a concern until post-touchdown telemetry and signal strength revealed that it had deployed during either retro fire or upon spacecraft contact with the lunar surface.

In planning the midcourse maneuver, a new aiming point (2.33 deg south, 43.83 deg west) was chosen about 1 deg farther north than the original target to improve the probability of landing in a smooth region. During visibility of DSS-11, Goldstone, a roll-yaw maneuver sequence was conducted in preparation for midcourse velocity correction. Then, at 06:45 GMT on May 31, 1966, a velocity correction of 20.35 m/sec was executed. Only a fraction of this velocity change was required to compensate for the injection error. The larger maneuver was selected to optimize the terminal descent conditions, which included main retro burnout velocity, vernier system propellant margin, and arrival time.

In preparation for terminal descent, a roll-yaw-roll maneuver sequence was initiated 38 min before retro ignition to properly align the retrorocket nozzle. This sequence was utilized in place of the normal two-maneuver sequence to optimize signal strength from Omniantenna B and to utilize proven attitude maneuvers.

The spacecraft altitude marking radar provided a *mark* signal that started the automatic descent sequence. After a delay of 7.826 sec, the three liquid-propellant vernier engines ignited, followed (after an additional 1.1-sec delay) by ignition of the solid-propellant main retro motor at a lunar slant range of 248,635 ft and velocity of 8,565 ft/sec. Main retro burnout occurred after the spacecraft had been decelerated to a velocity of 428 ft/sec. After the solid motor case was ejected, the radar altimeter and doppler velocity sensor (RADVS) became operational and controlled the final descent by throttling the vernier engines. The vernier engines were cut off when the spacecraft was 12 ft from the surface and traveling only 4 ft/sec. The spacecraft free-fell the final 12 ft, touching the surface at 06:17:37 GMT on June 2, 1966, with a velocity of 12 ft/sec. The three footpads contacted the surface nearly simultaneously (within 19 millisec of each other) and penetrated the surface to a depth of a few inches. At least one of the three crushable blocks under the frame also made an imprint on the surface as the shock absorbers compressed at touchdown. Strain-gage readings indicated that the spacecraft rebounded about 2½ in. above the surface before coming to rest.

The best estimate of the actual landing site, based upon preliminary analysis of post-touchdown tracking data, is −2.58 deg latitude and 316.65 deg longitude in the Ocean of Storms, about 9 miles from the modified aiming point. This result is in good agreement with landing site determinations discussed in Part II of this report, which are based upon correlation of *Surveyor I* topographical photos with earth-based pictures.

On a scale of miles, the spacecraft landed on a relatively smooth, nearly level, bare surface, encircled by hills and low mountains, the crestlines of a few of which are visible along the horizon from the vantage point of the spacecraft survey camera (about 4 ft above the surface). On a finer scale, the terrain surrounding *Surveyor I* is a gently rolling surface composed of granular material of a wide size range, studded with craters having diameters from a few inches to several hundred yards, and littered with fragmental debris ranging in size from less than 0.04 in. to more than a yard. The larger craters observed resemble those seen in the *Ranger* photographs in shape and distribution, leading to the conclusion that the *Surveyor I* site is representative of a mare surface.

Touchdown occurred 57 hr into the first lunar day (May 30–June 14), with about 5 hr remaining in the view period of DSS-11 at Goldstone, California. After postlanding engineering interrogations confirmed that all spacecraft systems had survived the landing, the first 200-scan-line picture was transmitted about 35 min after touchdown. After 13 additional 200-line pictures were received, the solar panel was positioned to receive maximum solar power, and the high-gain antenna was pointed toward the earth for transmission of higher quality 600-line pictures. Goldstone received 133 600-line pictures during the remainder of its first post-touchdown "pass," before it was necessary to command the spacecraft to an engineering telemetry mode for transfer to DSS-42 at Canberra, Australia.

By taking advantage of shading afforded by other elements of the spacecraft, it was possible to operate the spacecraft without exceeding the thermal limits during each remaining Goldstone pass of the first lunar day, except on June 8 and 9, when operations were suspended. In addition to a vast number of wide- and narrow-angle lunar surface pictures under varying lighting conditions, photographs were obtained of several bright stars in an effort to determine more accurately the location of the *Surveyor* spacecraft. (Because of the limited elevation range of the camera mirror, it was not possible to photograph the earth.) Pictures were also taken during and after the firing of bursts of nitrogen gas from the spacecraft attitude control jets in an unsuccessful attempt to disturb the lunar surface. A number of pictures were taken of the solar corona just after sunset. Over 10,000 pictures were received before the camera was secured after sunset of the first lunar day. The last of these pictures was successfully taken after the terminator had passed by increasing the exposure time to 4 min to permit sufficient earth-reflected light to illuminate the spacecraft footpad No. 2.

The spacecraft was in excellent condition as the first lunar day came to a close and, as a result, was secured in a configuration which was most favorable for its revival during the second lunar day.

Repeated attempts were made to communicate with the spacecraft beginning just before sunrise of the second lunar day, June 28. It finally responded on July 6, proving it had successfully survived the long, cold lunar night. The spacecraft solar panel was carefully commanded to a position to permit recharging of the battery, which had become very weak.

Shortly after resuming video operations on the second lunar day, a spacecraft battery anomaly occurred, causing an excessive rise in temperature. After emergency steps had been initiated to utilize what were believed to be the last few hours of spacecraft life, the battery returned to normal operation. It was then possible to continue taking picture sequences until the close of the second lunar day, July 14, when the operational phase of Mission A was terminated.

II. SPACE VEHICLE PREPARATIONS AND LAUNCH OPERATIONS

The *Surveyor* spacecraft was assembled and flight-acceptance-tested at Hughes Aircraft Corporation, El Segundo, California. On February 19, 1966, it was shipped to the Combined Systems Test Stand (CSTS) at San Diego for compatibility tests with the *Atlas/Centaur* (AC-10) launch vehicle. Immediately after completion of these tests the spacecraft and launch vehicle were air-lifted by C-133 aircraft to the Air Force Eastern Test Range (AFETR). Prelaunch assembly, checkout, and systems tests were carried out successfully, and launch was accomplished at the opening of the launch window on the first day of the scheduled launch period at 0941:00 EST, May 30, 1966.

A. Spacecraft Assembly and Testing

Initial assembly of *Surveyor I* was accomplished in the fall of 1964 and was followed by periods of Systems Group Testing, Initial Systems Checkout, Mission Sequence Tests, and some upgrade and rework periods. This sequence of tests was, in turn, followed by flight-acceptance vibration tests in June of 1965. The test sequence consisted of two lateral-axes vibrations and a Z-axis vibration.

In these initial system-level tests, numerous design problems were discovered and, after vibration testing, a period of spacecraft upgrade and a series of special tests were performed to improve the spacecraft performance. These tests were followed by a spacecraft checkout sequence early in August 1965 and prior to committing the spacecraft to Mission Sequence Tests in the solar-thermal-vacuum (STV) environment.

The solar-thermal-vacuum tests were conducted in the vacuum chamber at the HAC facility in El Segundo, starting in mid-August 1965. The tests consisted of two full-duration mission sequences at higher (112%) and lower (87%) than the expected Sun intensity, followed by a shorter, "plugs out" sequence at nominal Sun intensity. Considerable difficulty was experienced with spacecraft performance during the STV tests, resulting in either test aborts or tests not considered acceptable. After spacecraft rework and retest, the sixth attempt at the STV sequence resulted in acceptable spacecraft performance and was completed on January 22, 1966.

Because of the problems encountered in the STV sequences and resultant hardware rework and change, the

Z-axis vibration test was repeated late in January 1966. This test was followed by a vernier engine vibration test to verify proper radar altimeter and doppler velocity sensor (RADVS) performance under simulated retro engine burn and a test to establish proper flight control stability under dynamic conditions.

The spacecraft was then shipped by van to the Combined Systems Test (CST) facility in San Diego on February 19, 1966.

B. Combined Systems Test at San Diego

The Combined Systems Test between *Surveyor I* and the *Atlas/Centaur* (AC-10) launch vehicle was performed at the CSTS in San Diego. The prime purpose of the test was to verify physical compatibility of the interface between the *Centaur* adapter/nose fairing system and the spacecraft and to demonstrate operational compati-

bility between the launch vehicle and the spacecraft during a simulated countdown and flight. The configuration for the test is shown in Fig. II-1. The *Atlas* was installed in a horizontal test position on December 29, 1965, followed by erection of the *Centaur* and interstage adapter in the adjoining vertical test stand on January 31, 1966.

After arrival at the CSTS on February 19, 1966, the spacecraft was reassembled and checked out with no problems noted. Mating of the spacecraft with the forward adapter, encapsulation with the nose fairing, and mating with the *Centaur* were accomplished on March 4, 1966. Only minor problems were noted with the spacecraft separation hardware and interface electrical connector orientation. After the spacecraft was in place on the *Centaur*, a Systems Readiness Test (SRT) and practice countdown were performed by the spacecraft on March 5, 1966.

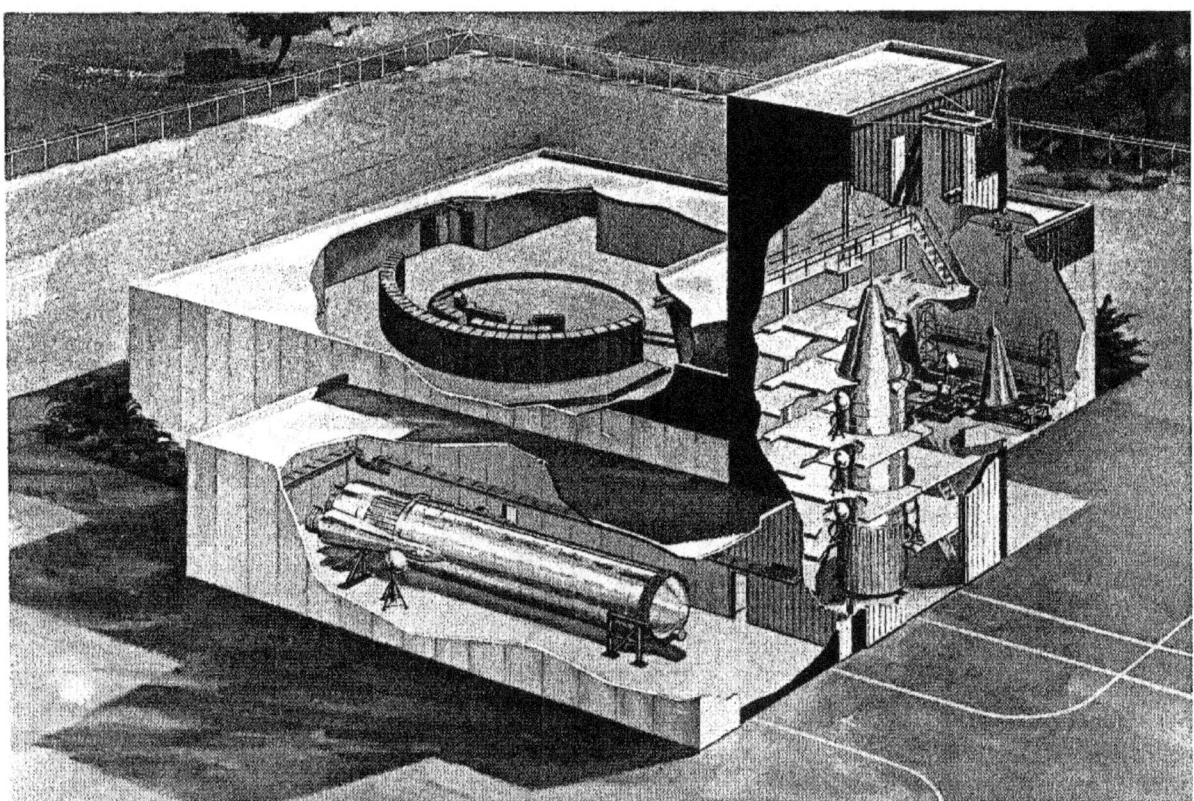

Fig. II-1. Combined System Test Stand (CSTS) at San Diego

The Combined Systems Test was performed on March 7, 1966. Data analysis indicated only minor interface problems, and the spacecraft was demated from the *Centaur* on March 8, 1966. After preparation for shipment, the spacecraft was transported by air from Miramar, San Diego, to AFETR on March 13, 1966. *Atlas* and *Centaur* shipment, also by air, occurred, respectively, March 14 and 16.

C. Launch Operations at AFETR

The major operations performed at AFETR after arrival of the spacecraft, the spacecraft support equipment, and the launch vehicle are listed in Table II-1.

1. Initial Preparations

Following inspection of the spacecraft and reassembly after arrival at Hangar AO on March 14, 1966, a series of performance verification tests was performed. These initial spacecraft tests were completed by April 5, 1966. Included in these tests were calibrations of the survey television camera. No significant spacecraft problems were noted.

The spacecraft then was moved to the Explosive Safe Facility (ESF) and prepared for the first trip to the launch pad. Preparations included (1) flight-level pressurization of the flight control gas system so that a leak test could be accomplished during the on-pad period, (2) installation on the forward adapter, and (3) encapsulation of the spacecraft. During this period, the battery charge regulator (BCR) was damaged by a short in the main battery harness. A temporary substitution was made

and the damaged BCR was repaired and replaced after the Joint Flight Acceptance Composite Test (J-FACT). The spacecraft was moved to the launch pad and mated with the *Centaur* on April 17, 1966.

2. Propellant Tanking and Flight Acceptance Composite Tests

After the spacecraft was mated with the *Centaur*, a series of tests was performed to verify proper operation with the blockhouse equipment and the RF air link between the launch pad and Hangar AO. All control of the spacecraft except for external power and some monitor functions is at Hangar AO. On April 17 and 18, an SRT and practice countdown were performed, and on April 19 a compatibility test between the spacecraft and the DSS 71 station at Cape Kennedy was successfully accomplished.

The spacecraft next participated in the launch vehicle Propellant Tanking Test. During this test the launch pad stand is moved back and the complete vehicle is exposed to off- and on-board RF sources as in the launch countdown. No RF interference problems were noted.

The Joint Flight Acceptance Composite Test (J-FACT) was performed on 26 April. This test, involving all systems, included a simulated countdown followed by a simulated launch through injection with all umbilicals removed. After completion of the test, the spacecraft was demated and transported to the ESF. One major problem occurred when the *Atlas* programmer was switched from internal to external power, as planned, during a "hold

Table II-1. Major operations at Cape Kennedy

Operation	Location	Date completed, 1966
AC-10 erection	Launch Complex 36A	March 31
SC-I (*Surveyor I*) inspection, reassembly, and initial testing	Hangar AO	April 5
SC-I preparations for on-pad testing; encapsulation	Explosive Safe Facility	April 15
SC-I mate to *Centaur*	Launch Complex 36A	April 17
Propellant Tanking Test	Launch Complex 36A	April 20
Joint Flight Acceptance Composite Test	Launch Complex 36A	April 26
SC-I demated	Launch Complex 36A	April 26
SC-I decapsulation and alignment checks	Explosive Safe Facility	May 3
SC-I RADVS test and system checkout	Hangar AO	May 14
Flight Acceptance Composite Test (without SC-I)	Launch Complex 36A	May 18
SC-I propellant loading, pressurization, and weight and balance checks	Explosive Safe Facility	May 24
SC-I remated to *Centaur*	Launch Complex 36A	May 26
Launch	Launch Complex 36A	May 30

fire" test which had been incorporated into the last few seconds of the automatic countdown sequence. This event caused loss of programmer liftoff reference time and scrambling of the pitch program switch positions. The failure required removal and replacement of the *Atlas* programmer and minor design modification to the flight and spare units. The *Centaur* programmer was also replaced at this time.

After replacement of the *Atlas* and *Centaur* programmers, a second FACT was conducted on May 18 without the spacecraft. No anomalies were encountered and the results of this test were satisfactory.

3. Final Flight Preparations

Following the J-FACT of April 26, the spacecraft was decapsulated and depressurized in the ESF, and alignment checks were completed on May 3, 1966. The spacecraft then was moved to Hangar AO, where final flight preparations were continued. The preparations at Hangar AO included a ranging test of the RADVS and other system-level checkouts to establish proper operation prior to launch. The only significant problem during this test phase was the failure of the RADVS signal data converter. This unit was replaced with a flight spare.

The spacecraft was transported to the propellant loading building at the ESF on May 14, 1966. After two days of propellant loading, the spacecraft was moved to the assembly laboratory at the ESF for weight and balance operations and alignment checks. After installation of the flight retrorocket, a recheck was made of some alignments and the spacecraft was mated to the forward adapter. On May 24, 1966, flight pressurization of both the flight control and vernier propulsion systems was completed, followed by final electrical checks and encapsulation with the nose fairing.

Significant problems during these final preparations included:

1. Replacement of the main battery and monitor cable because of a short circuit in the cable and possible overstress of the battery.

2. Replacement of the roll drive assembly in the antenna solar panel positioner (ASPP) due to excessive friction.

After a System Readiness Test on May 25, the spacecraft was moved to the launch pad and mated early on May 26, 1966. No-voltage tests, through the spacecraft umbilical connections were followed by an SRT and practice countdown to assure readiness for the launch countdown.

Several field changes were made to the *Atlas/Centaur* following the May 18 FACT. These changes, which developed from a review of AC-8 flight test results, included:

1. Installation of improved 3.5- and 6-lb hydrogen peroxide engines.

2. Installation of redesigned *Centaur* insulation panel hinges.

3. Incorporation of additional insulation panel break-wire instrumentation for monitoring the panel jettison sequence.

In addition, the spacecraft inadvertent separation, automatic destruct switch circuitry was disabled after Range Safety Office approval had been obtained. However, ground command destruct capability was retained.

4. Countdown and Launch

The final spacecraft SRT test was started at 2200 EST, May 29, at a countdown time of $T-550$ min. The SRT was completed well within the time allotted, at 0352 EST May 30, and test personnel were given a two-hour break prior to joining the launch vehicle countdown at 0700 EST.

The mission was scheduled for a 381-min countdown, starting at 0320 EST, May 30. The countdown proceeded without interruption to $T-90$ min (0650 EST), when a scheduled 1-hr hold was called. The spacecraft system entered the joint countdown during the $T-90$ hold. The countdown was resumed at 0750 EST and proceeded as planned to $T-5$ min, when a second built-in hold of 21 min was started. The countdown was resumed again at 0936 and proceeded smoothly through liftoff ($T-0$). Liftoff occurred (Fig. II-2) at 0941:00.990 EST (1441:00:990 GMT).

The built-in holds had been scheduled into the countdown to increase the probability of launch-on-time. Any unscheduled holds could be subtracted from these planned hold times; however, there were no unscheduled holds on the AC-10/SC-I launch.

The general performance of mechanical ground support equipment was satisfactory throughout the launch countdown. Minor difficulties were encountered but these had no detrimental effect on the space vehicle system.

All systems performed normally throughout the launch, and the spacecraft was injected into a satisfactory lunar trajectory. The powered-flight sequence of events and launch vehicle performance are described in Section III.

The atmospheric conditions on launch day were favorable. Surface winds were 7 knots from 240 degrees, with unrestricted visibility of 10 miles. Surface temperature was 82°F with relative humidity of 67%. Cloud cover consisted of 0.4 cumulus at 2200 ft and 0.1 alto stratus at 10,000 ft. Projected winds aloft were reported as 57 knots at 43,000 ft, with a maximum expected shear parameter of 7 knots per thousand feet of altitude.

D. Launch Phase Real-Time Mission Analysis

1. Countdown to Launch

The launch windows which were established for the May/June 1966 launch period are shown in Fig. II-3. During countdown operations, those factors acting to constrain the launch window or period were continually evaluated by the Launch Phase Mission Analyst. The Mission Director was advised of these evaluations for consideration in the launch or hold decision.

At $T-154$ min, RF propagation problems interfered with the retransmission of tracking and telemetry data from the Range Instrumentation Ship (RIS) *Arnold*. The Mission Director was advised that an adequate quantity of tracking data was available from Antigua for the AFETR/RTCS to compute DSS-51 acquisition information (see Fig. II-4). He was also advised that in the event that Antigua data was lost, the tracking data analysts in FPAC would accept DSS-51 predicts based on Ascension's post-retro tracking data. In addition, it was agreed that the voice readouts of ten selected spacecraft telemetry parameters would be acceptable in lieu of real-time transmission of the spacecraft telemetry should RF propagation cause further difficulty.

By 40 min into the 60-min built-in hold at $T-90$ min, both the tracking and telemetry data retransmission problems had cleared up. At $T-72$ min, an electrical storm between San Salvadore and Grand Turk interrupted both the HF and subcable tracking data transmission links from key downrange stations. However, telemetry data transmission was still good. Attempts were made to determine whether such conditions would cause the Range Safety Officer (RSO) to call a hold. Although definite answers to these inquiries were not obtained, indications were that the RSO would accept these conditions and rely on the Deputy RSO's stationed at downrange stations.

Fig. II-2. *Atlas/Centaur AC-10 lifting off with Surveyor I*

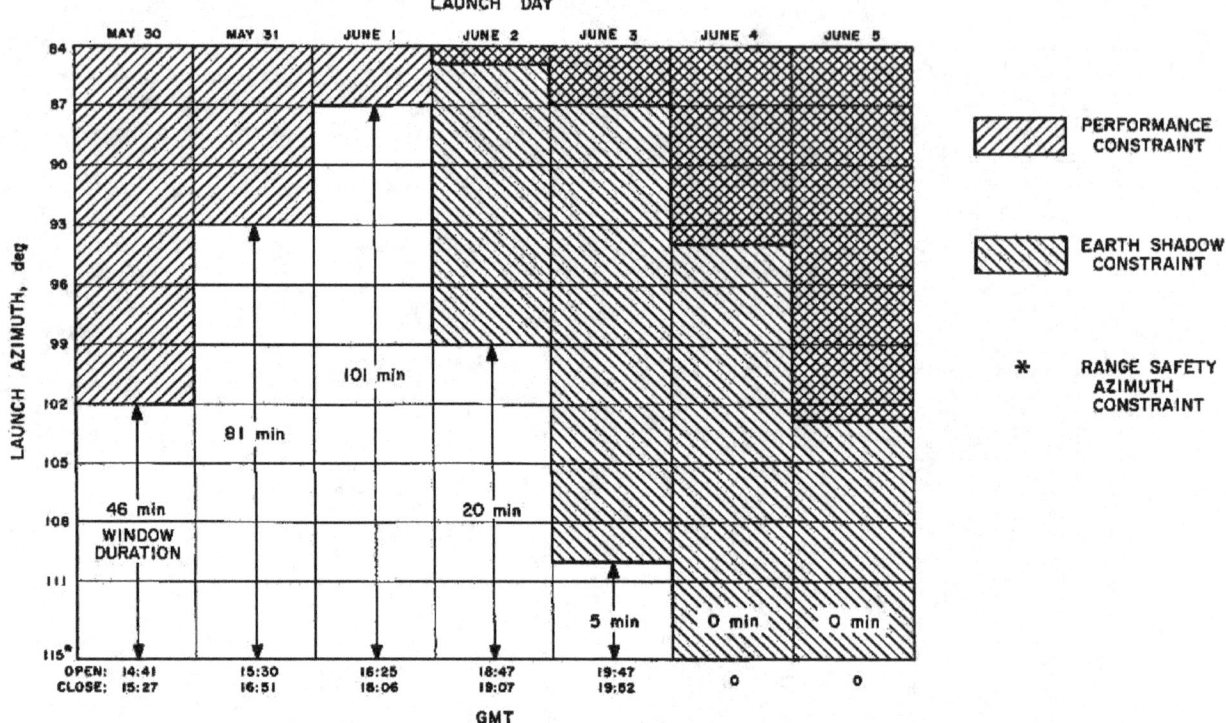

Fig. II-3. Surveyor launch window design for May/June 1966 launch window

During the 21-min built-in hold at $T-5$ min, the tracking data transmission problems cleared up.

At $T-60$ sec, both digital trackers on the RIS *Arnold* experienced failures. Based on the earlier evaluations and recommendations the countdown was permitted to proceed to launch.

2. Launch to DSN Acquisition

During the launch-to-DSN-acquisition phase of the flight, all available information was obtained on the performance of the space vehicle and associated flight operations in order to provide the Mission Director and Mission Operations Staff with a composite status of the mission. This continually updated status was utilized to ascertain any possible influence on the planned DSIF acquisitions and mission operations.

The nominal powered flight of the launch vehicle precluded any necessity for a significant amount of near-earth mission analyses. However, since the initial pre-(Centaur) retro orbit calculation by the RTCS yielded a poor orbit, it became necessary to depend solely on the tracking station acquisition times and launch vehicle telemetry for confirmation of a nominal flight. Furthermore, no confidence was placed on the quality of the initial orbit since Antigua lost track so early that only 11 points of ambiguous tracking data were obtained, and the RIS *Arnold* provided no tracking data. The likelihood of a nominal flight became quite high when DSS-51 acquired the spacecraft within seconds of the nominal time.

When indications were received that Omniantenna A had probably not extended, launch vehicle telemetry was analyzed for the remote chance that the spacecraft anomaly could have been caused by a launch vehicle event. No such evidence was found.

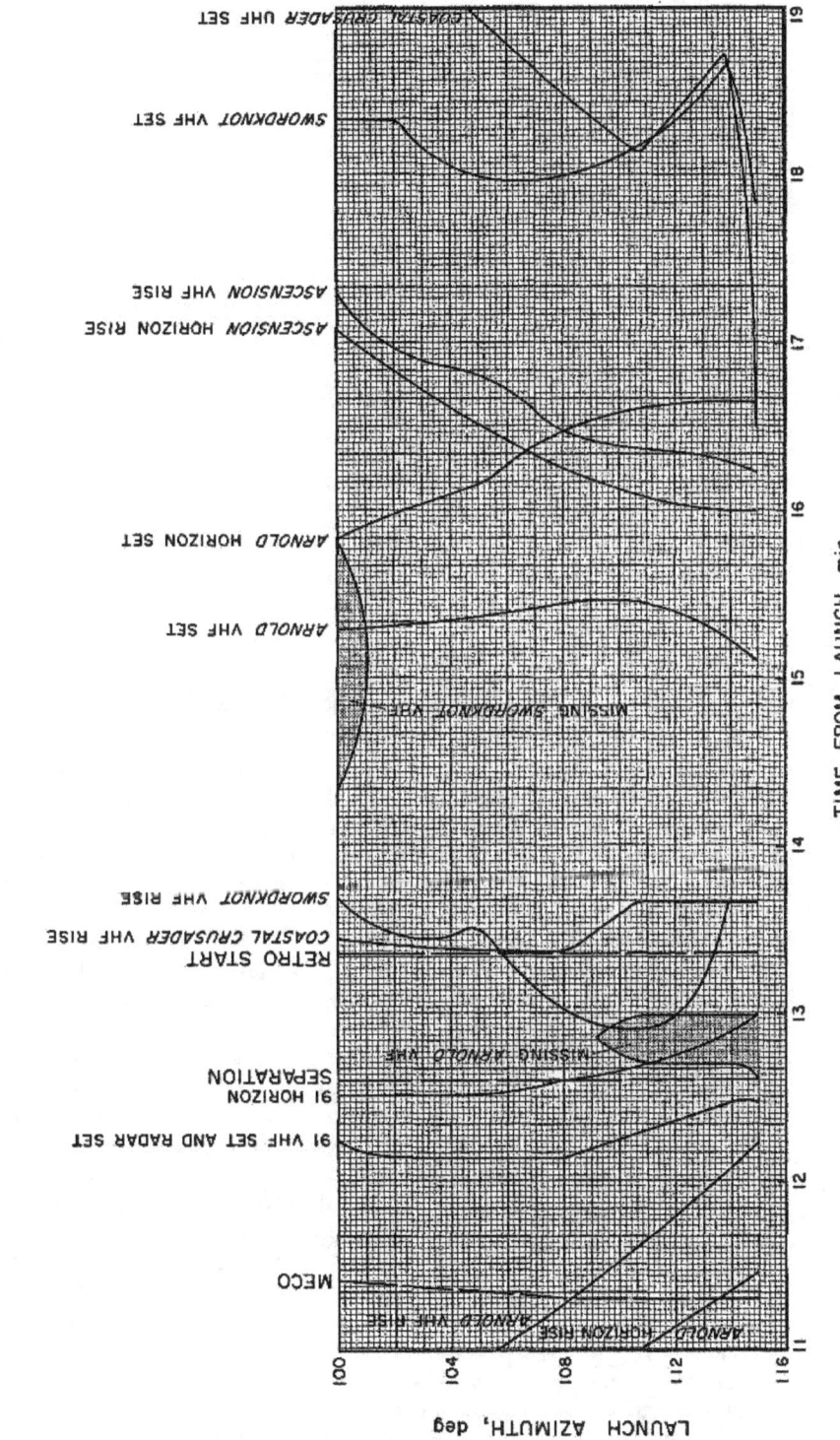

Fig. II-4. Station coverage capability, Surveyor Mission A

III. LAUNCH VEHICLE SYSTEM

The *Surveyor* spacecraft was injected into its lunar transit trajectory by a General Dynamics *Atlas/Centaur* vehicle (AC-10). The vehicle was launched on a "direct ascent" powered flight from Launch Complex 36A of the AFETR at Cape Kennedy, Florida. This mission marked the first operational flight of an *Atlas/Centaur* vehicle. A total of seven vehicle R&D flight tests had been performed or attempted over the preceding four years. Operational capability for "direct ascent" missions was demonstrated with the successful flight test of AC-6 on August 11, 1965. One additional *Centaur* dual-burn development flight is scheduled to demonstrate capability to launch via "parking orbit" ascent trajectories.

The *Atlas/Centaur* vehicle with the *Surveyor* spacecraft encapsulated in the nose fairing is 113 ft long and weighs 302,750 lb at liftoff (2-in. rise). The basic diameter of the vehicle is a constant 10 ft from the aft end to the base of the conical section of the nose fairing. The configuration of the completely assembled vehicle is illustrated in Fig. III-1. Both the *Atlas* first stage and *Centaur* second stage utilize thin-wall, pressurized, main propellant tank sections of monocoque construction to provide primary structural integrity and support for all vehicle systems. The first and second stages are joined by an interstage adapter section of conventional sheet

and stringer design. The clamshell nose fairing is constructed of laminated fiberglass over a fiberglass honeycomb core and attaches to the forward end of the *Centaur* cylindrical tank section.

A. Atlas Stage

The first stage of the *Atlas/Centaur* vehicle is a modified version of the *Atlas D* used on many previous NASA and Air Force missions such as *Ranger*, *Mariner* and *OGO*. The *Atlas* propulsion system burns RP-1 kerosene and liquid oxygen in each of its five engines to provide a total liftoff thrust of approximately 388,000 lb. The individual sea-level thrust ratings of the engines are: two booster engines at 165,000 lb each; one sustainer engine at 57,000 lb; and two vernier engines at 670 lb each. The *Atlas* can be considered a 1½-stage vehicle because the "booster section," weighing 6000 lb and consisting of the two booster engines together with the booster turbo pumps and other equipment located in the aft section, is jettisoned after about 2.5 min of flight. The sustainer and vernier engines continue to burn until propellant depletion. A mercury manometer propellant utilization system is used to control mixture ratio for the purpose of minimizing propellant residuals at *Atlas* burnout.

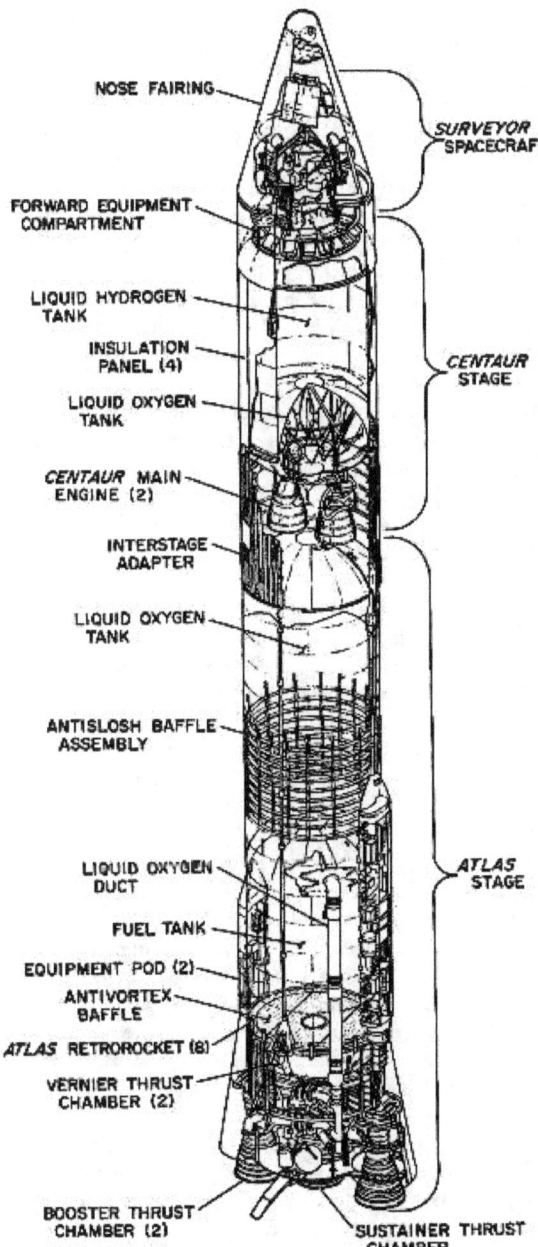

NOSE FAIRING

SURVEYOR SPACECRAFT

FORWARD EQUIPMENT COMPARTMENT

LIQUID HYDROGEN TANK

INSULATION PANEL (4)

LIQUID OXYGEN TANK

CENTAUR STAGE

CENTAUR MAIN ENGINE (2)

INTERSTAGE ADAPTER

LIQUID OXYGEN TANK

ANTISLOSH BAFFLE ASSEMBLY

ATLAS STAGE

LIQUID OXYGEN DUCT

FUEL TANK

EQUIPMENT POD (2)

ANTIVORTEX BAFFLE

ATLAS RETROROCKET (8)

VERNIER THRUST CHAMBER (2)

BOOSTER THRUST CHAMBER (2)

SUSTAINER THRUST CHAMBER

Fig. III-1. Atlas/Centaur/Surveyor space vehicle configuration

Flight control of the first stage is accomplished by the *Atlas* autopilot, which contains displacement gyros for attitude reference, rate gyros for response damping, and

a programmer to control flight sequencing until *Atlas/Centaur* separation. After booster jettison, the *Atlas* autopilot also is fed steering commands from the all-inertial guidance set located in the *Centaur* stage. Vehicle attitude and steering control are achieved by the coordinated gimballing of the five thrust chambers in response to autopilot signals.

The *Atlas* contains a single VHF telemetry system which transmits data on 118 first-stage measurements until *Atlas* separation. The system operates on a frequency of 229.9 mc over two antennas mounted on opposite sides of the vehicle at the forward ends of the equipment pods. Redundant range-safety command receivers and a single destructor unit are employed on the *Atlas* to provide the Range Safety Officer with means of terminating the flight by initiating engine cutoff and destroying the vehicle. The system is inactive after normal *Atlas* staging occurs. The AZUSA tracking system has been deleted from the *Atlas* for *Surveyor* missions, leaving only the C-band tracking system on the *Centaur* stage.

B. Centaur Stage

The *Centaur* second stage is the first vehicle to utilize liquid hydrogen/liquid oxygen, high-specific-impulse propellants. The cryogenic propellants require special insulation to be used for the forward, aft, and intermediate bulkheads as well as the cylindrical walls of the tanks. The cylindrical tank section is thermally insulated by four jettisonable insulation panels having built-in fairings to accommodate antennas, conduits, and other tank protrusions. The insulation panel hinges were redesigned for AC-10 to overcome a deployment control problem which had been suspected on vehicle development flights AC-6 and AC-8. Most of the *Centaur* electronic equipment packages are mounted on the forward tank bulkhead in a compartment which is air-conditioned prior to liftoff.

The *Centaur* is powered by two constant-thrust engines rated at 15,000 lb thrust each in vacuum. Each engine can be gimballed to provide control in pitch, yaw, and roll. Propellant is fed from each of the tanks to the engines by boost pumps driven with hydrogen peroxide turbines. In addition, each engine contains integral "bootstrap" pumps driven by hydrogen propellant, which is also used for regenerative cooling of the thrust chambers. A propellant utilization system is used on the *Centaur* stage to achieve minimum residual of one propellant at depletion of the other. The system controls the mixture ratio valves as a continuous function of propellant in the tanks by means of tank probes and an error ratio detector. The nominal oxygen/hydrogen mixture ratio is 5:1 by weight.

The second-stage all-inertial guidance system contains an on-board computer which provides vehicle steering commands after jettison of the *Atlas* booster section. The *Centaur* guidance signals are fed to the *Atlas* autopilot until *Atlas* sustainer engine cutoff and to the *Centaur* autopilot after *Centaur* main engine ignition. This mission was the first *Centaur* flight to employ an inertial platform containing new gyros having reduced gimbal stop angles, improved flex leads, better balanced spin motor, and reduced synchronous torque sensitivity. It was also the first flight during which the gyros were not torqued to correct for gyro drift characteristics. Gyro drifts were compensated for by the guidance system computer, which was programmed to set the torquing signals to zero during flight. The *Centaur* autopilot system provides the primary control functions required for vehicle stabilization during powered flight, execution of guidance system steering commands, and attitude orientation following the powered phase of flight. In addition, the autopilot system employs an electromechanical timer to control the sequence of programmed events during the *Centaur* phase of flight, including a series of commands required to be sent to the spacecraft prior to spacecraft separation.

The *Centaur* reaction control system provides thrust to control the vehicle after powered flight. For small corrections in yaw, pitch, and roll attitude control, the system utilizes six individually controlled, fixed-axes, constant-thrust, hydrogen peroxide reaction engines. These engines are mounted in clusters of three, 180 deg apart on the periphery of the main propellant tanks at the interstage adapter separation plane. Each cluster contains one 6-lb-thrust engine for pitch control and two 3.5-lb-thrust engines for yaw and roll control. In addition, four 50-lb-thrust hydrogen peroxide engines are installed on the aft bulkhead, with thrust axes parallel with the vehicle axis. These engines are for use during retromaneuver and for executing larger attitude corrections if necessary. The cluster engines were slightly modified from the design used on the previous flight (AC-8) in that the large aluminum B-nut on the thrust chambers was replaced with a steel flange joint to effect a more positive seal.

The *Centaur* stage utilizes a VHF telemetry system with a single antenna transmitting through the nose fairing cylindrical section on a frequency of 225.7 mc. The telemetry system provides data on 140 measurements from transducers located throughout the second stage

and spacecraft interface area as well as a spacecraft composite signal from the spacecraft central signal processor.

Redundant range safety command receivers are employed on the *Centaur*, together with shaped charge destruct units for the second stage and spacecraft. This provides the Range Safety Officer with means to terminate the flight by initiating *Centaur* main engine cutoff and destroying the vehicle and spacecraft retrorocket. The system can be safed by ground command, which is normally transmitted by the Range Safety Officer when the vehicle has reached injection energy.

Prior to final encapsulation and mating of *Surveyor*, a waiver was obtained to permit the disabling of an inadvertent separation system on Mission A, which would have provided for the automatic destruction of the *Centaur* and spacecraft in the event of premature spacecraft separation.

A C-band tracking system is contained on the *Centaur* which includes a lightweight transponder, circulator, power divider, and two antennas located under the insulation panels. The C-band radar transponder provides real-time position and velocity data for the Range Safety Instantaneous Impact Predictor as well as data for use in guidance and trajectory analysis.

C. Launch Vehicle/Spacecraft Interface

The general arrangement of the *Surveyor/Centaur* interface is illustrated in Fig. III-2. The spacecraft is completely encapsulated within a nose fairing/adapter system in the final assembly bay of the Explosive Safe Facility prior to being moved to the launch pad. This encapsulation provides protection for the spacecraft from the environment before launch as well as from aerodynamic loads during ascent.

The spacecraft is first attached to the forward section of a two-piece, conical adapter system of aluminum sheet and stringer design by means of three latch mechanisms, each containing a dual-squib pin puller. The following equipment is located on the forward adapter: three separation spring assemblies each containing a linear potentiometer for monitoring separation; a 52-pin electrical connector with a pyrotechnic separation mechanism; three pedestals for the spacecraft-mounted separation sensing and arming devices; a shaped-charge destruct assembly directed toward the spacecraft retromotor; an accelerometer for monitoring lateral vibration at the separation

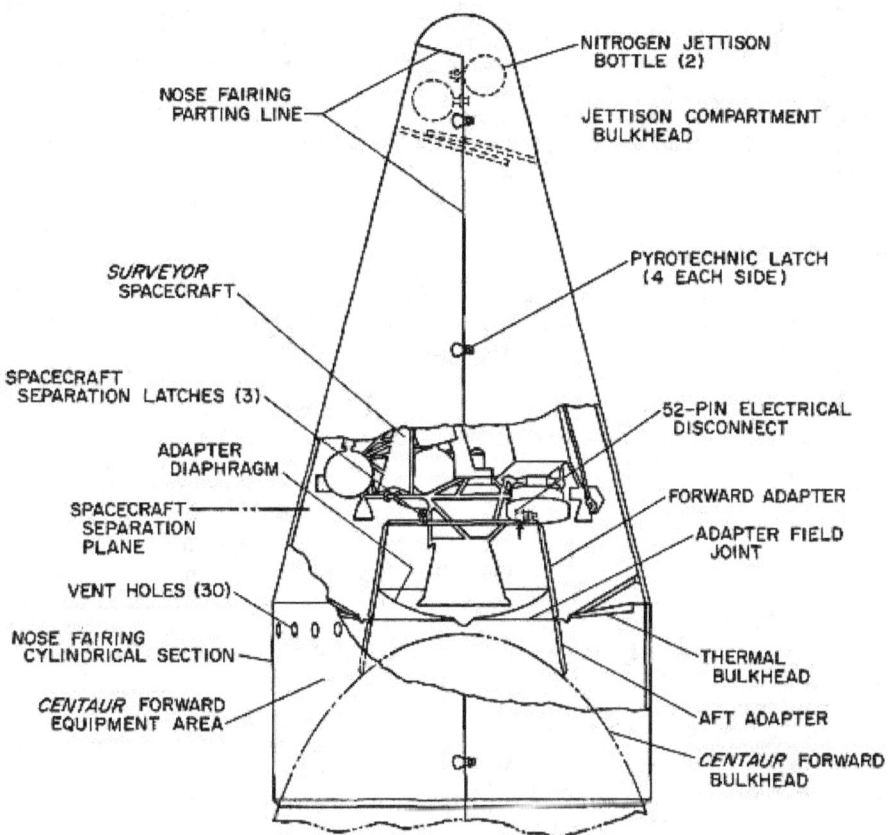

Fig. III-2. Surveyor/Centaur interface configuration

plane; and a diaphragm to provide a thermal seal and to prevent contamination from passing to the spacecraft compartment from the *Centaur* forward equipment compartment.

The low-drag nose fairing is an RF-transparent, clamshell configuration consisting of four sections fabricated of laminated fiberglass cloth faces and honeycomb fiberglass core material. Two half-cone forward sections are brought together over the spacecraft mounted on the forward adapter. An annular thermal bulkhead between the adapter and base of the conical section completes encapsulation of the spacecraft.

The encapsulated spacecraft assembly is mated to the *Centaur* at a flange field joint requiring 72 bolts between the forward and aft adapter sections. The remaining two half-cylindrical sections of the nose fairing are attached to the forward end of the *Centaur* tank around the equip-

ment compartment prior to mating of the spacecraft. Doors in the cylindrical sections provide access to the adapter field joint. The electrical leads from the forward adapter are carried through three field connectors and routed across the aft adapter to the *Centaur* umbilical connectors and to the *Centaur* programmer and telemetry units.

Special distribution ducts are built into the nose fairing and forward adapter to provide air conditioning of the spacecraft cavity after encapsulation and until liftoff. Seals are provided at the joints to prevent shroud leakage except out through vent holes in the cylindrical section.

The entire nose fairing is designed to be ejected by separation of two clamshell pieces, each consisting of a conical and cylindrical section. Four pyrotechnic pin-puller latches are used on each side of the nose fairing to carry the tension loads between the fairing halves. A bolted connection, with a flexible linear-shaped charge for separation, transmits loads between the nose fairing and *Centaur* tank. A nitrogen bottle is mounted in each half of the nose fairing near the forward end to supply gas for cold gas jets to force the panels apart. Hinge fittings are located at the base of each fairing half to control ejection, which occurs under vehicle acceleration of approximately 1 *g*.

D. Vehicle Flight Sequence of Events

All vehicle flight events occurred as programmed at near nominal times with no anomalies. Predicted and actual times for the vehicle flight sequence of events are included in Table A-1 of Appendix A. Following is a brief description of the vehicle flight sequence of events with all times referenced to liftoff (2 in. rise) unless otherwise noted.

1. *Atlas* Booster Phase of Flight

Hypergolic ignition of all five *Atlas* engines was initiated 2 sec before liftoff, with liftoff occurring at 14:41:00.990 GMT. The launcher mechanism is designed to begin a controlled release of the vehicle when all engines have reached nearly full thrust. At 2 sec after liftoff, the vehicle began a 13-sec programmed roll from the fixed launcher azimuth setting of 105 deg to the desired launch azimuth of 102.285 deg. The programmed pitchover of the vehicle began 15 sec after liftoff and lasted until booster engine cutoff (BECO).

The vehicle reached Mach 1 at 58 sec and maximum aerodynamic loading occurred at 77 sec. During the booster phase of flight the booster engines were gimballed for pitch, yaw, and roll control, and the vernier engines were active in roll control only while the sustainer engine was centered.

At 142.0 sec, BECO was initiated by a signal from the *Centaur* guidance system when vehicle acceleration equalled 5.68 *g* (expected value: 5.7 ±0.08 *g*). At 3.1 sec after BECO, with the booster and sustainer engines centered, the booster section was jettisoned by release of pneumatically operated latches.

2. *Atlas* Sustainer Phase of Flight

At BECO+8 sec the *Centaur* guidance system was enabled to provide steering commands for the *Atlas* sustainer phase of flight. During this phase the sustainer engine was gimballed for pitch and yaw control, while

the verniers were active in roll. The *Centaur* insulation panels were jettisoned by firing shaped charges at 175.8 sec at an altitude of approximately 49 nm where the aerodynamic heating rate was rapidly decreasing. At 202.8 sec, squibs were fired to unlatch the clamshell nose fairing, which was jettisoned 1.0 sec later by means of nitrogen gas thruster jets activated by pyrotechnic valves.

Other programmed events which occurred during the sustainer phase of flight were: the unlocking of the *Centaur* hydrogen-tank vent valve to permit venting as required to relieve hydrogen boiloff pressure; starting of the *Centaur* boost pumps 46 sec prior to *Centaur* main engine ignition (MEIG); and locking of the *Centaur* oxygen-tank vent valve followed by oxygen-tank pressurization.

Sustainer and vernier engine cutoff (SECO and VECO) occurred at 239.4 sec as a result of oxidizer propellant depletion with fuel depletion imminent. Shutdown began with a slow thrust decay due to oxidizer depletion. Final fast shutdown was initiated by closure of a switch when fuel manifold pressure dropped to 625 psi as a result of reduction in speed of the turbopump which also utilizes the main fuel and oxidizer propellants.

Separation of the Atlas from the *Centaur* occurred 1.9 sec after SECO by firing of shaped charges at the forward end of the interstage adapter. This was followed by ignition of eight retrorockets located at the aft end of the *Atlas* tank section to back the *Atlas*, together with the interstage adapter, away from the *Centaur*.

3. *Centaur* Phase of Flight Through Spacecraft Separation

The *Centaur* main engines were ignited 9.6 sec after *Atlas/Centaur* separation and burned for 438.4 sec, or until 689.2 sec. Main engine cutoff (MECO) was commanded by the guidance system when the desired injection conditions were reached. At main engine cutoff, the hydrogen peroxide engines were enabled for attitude stabilization.

During the 67.7-sec period between MECO and spacecraft separation, the following signals were transmitted to the spacecraft from the *Centaur* programmer: extend spacecraft landing gear; unlock spacecraft omniantennas; turn on spacecraft transmitter high power. An arming signal also was provided by the *Centaur* during this period to enable the spacecraft to act on the preseparation commands.

The *Centaur* commanded separation of the spacecraft electrical disconnect 5.5 sec before spacecraft separation, which was initiated at 756.9 sec. The *Centaur* attitude-control engines were disabled for 5 sec during spacecraft separation in order to minimize vehicle turning moments.

4. *Centaur* Retromaneuver

At 5 sec after spacecraft separation, the *Centaur* began a turnaround maneuver using the attitude-control engines to point the aft end of the stage in the direction of the flight path. Approximately midway in the turn and while continuing the turn, two of the 50-lb-thrust hydrogen peroxide engines were fired for a period of 20 sec to provide initial lateral separation of the *Centaur* from the spacecraft. At 240 sec after separation, the propellant blowdown phase of the *Centaur* retromaneuver was initiated by opening the hydrogen and oxygen prestart valves. Oxygen was vented through the engine nozzles while hydrogen discharged through vent tubes. Propellant blowdown was terminated after 250 sec by closing the prestart valves. At the same time (1246.9 sec) the *Centaur* power change-over switch was energized to turn off all power except telemetry and C-band beacon.

E. Performance

The close correlation of the actual and predicted flight path and events (see Appendix A) for the powered flight phase through injection was a result of the near nominal performance of the *Atlas/Centaur* vehicle system.

1. Guidance and Flight Control

Performance of the guidance system was satisfactory, as evidenced by the projected miss distance of the injected spacecraft of only about 250 nm from the prelaunch aiming point. This error was equivalent to about 1-sigma guidance system dispersions (root-sum-squared). However, the major factor contributing to target miss appears to be higher actual *Centaur* main-engines shutdown impulse than that utilized for "targeting" the guidance equations. Taking the shutdown impulse error into consideration, the remaining error due to guidance system hardware and software is extremely small.

The guidance system discrete commands (BECO, SECO backup, and MECO) were issued well within system tolerance. When guidance steering was admitted to the *Atlas* autopilot at BECO+8 sec, the vehicle was about 18 deg nose down and 1 deg nose right from the desired steering vector. This steering error was nulled in

approximately 11 sec. All guidance steering commands required during flight were easily handled by the system.

Autopilot performance was satisfactory throughout the flight. A BECO transient which excited the first bending mode in the pitch plane and resulted in large engine gimbal angles was damped out prior to booster jettison. No unusual disturbances were noted during the remainder of the flight. Pitch, yaw, and roll rate gyro outputs at the time of *Surveyor* separation indicated a smooth separation.

The *Centaur* reaction control system apparently performed properly. Hydrogen peroxide engine firing commands were not telemetered and firing times could only be approximated from limited engine temperature and vehicle rate gyro data; however, it was indicated that vehicle control was maintained throughout the entire post-MECO and retromaneuver period when the system was active. At power shutdown, vehicle rates were less than 0.05 deg/sec.

After *Atlas* separation, the *Centaur* is only roll-rate stabilized. Telemetry data indicates that the vehicle did not roll appreciably during the powered phase (less than 15 deg clockwise). However, as expected, the *Centaur* turnaround maneuver resulted in an approximately 90-deg inertial roll.

2. Propulsion and Propellant Utilization

Both *Atlas* and *Centaur* propulsion systems operated normally throughout the flight, although the *Centaur* main engines burned approximately 5 sec longer than predicted. The longer burn time was well within the allowable system dispersion and was apparently due to lower actual main engine thrust level than that used in preflight simulation.

All vehicle propellant systems performed satisfactorily. The *Atlas* propellant utilization (PU) system controlled propellants to effect nearly simultaneous depletion at SECO with liquid oxygen depletion shutdown. This resulted in minimum *Atlas* "burnable" residuals. However, unburnable liquid oxygen residuals remaining in the *Atlas* were greater than predicted in preflight trajectory simulation. In other words, sustainer engine shutdown contributed less impulse than is assumed in the current sustainer engine shutdown model.

The *Centaur* PU system also performed well, controlling the calculated unbalance of propellants at MECO to only 5 lb of liquid hydrogen. The "burnable" residuals were calculated to be 145 ±47 lb and would have permitted an additional burn time of 1.6 ±0.7 sec at normal engine flow rates until liquid oxygen depletion. The predicted value for burnable residuals was 195 lb for a completely nominal flight. The lower actual residuals may be attributed to lower than nominal engine performance and/or tanking levels.

3. Pneumatics, Hydraulics, and Electrical Power

Pressure stability and regulation were maintained within desired limits in both the *Atlas* and *Centaur* hydraulic and pneumatic circuits.

Performance of vehicle electrical power systems including range safety power supplies was normal throughout the flight.

4. Tracking, Telemetry, and Range Safety

Tracking (C-band) was obtained through *Centaur* main engine cutoff. However, several dropouts occurred beyond this point (see Section V).

A range safety command to disable the destruct systems was sent from Antigua 10.9 sec after MECO and was properly received and executed. The *Surveyor* inadvertent separation destruct system requirement had been waived by Range Safety for this mission and was inactive.

The *Atlas* and *Centaur* instrumentation and telemetry systems functioned satisfactorily, although some data is questionable. A temperature transducer on one of the hydrogen peroxide attitude control engine nozzles indicated unexpected trends which are not fully explained.

5. Vehicle Loads and Environment

Vehicle bending moments were less than predicted and within vehicle capability. The longitudinal load factor buildup was as expected, and a maximum value of 5.68 g was reached at BECO.

The vibration profile of the space vehicle was similar to those of preceding development flights. Throughout the *Centaur* program, longitudinal and lateral loads of significant magnitudes have been imparted to the vehicle during the initial launch sequence because of the launcher holddown and release pin locking and retracting mechanism. Vibrations measured in the vicinity of the spacecraft separation plane at liftoff and BECO appeared to be slightly greater than for *Centaur* flight AC-6, while those encountered at other events were comparable. The

low-frequency launcher release transient decayed by 15 sec after liftoff. Maximum aerodynamic loads occurred at liftoff +77 sec, with vibration levels lower than at launch. An unusually high first-bending-mode oscillation appeared in the pitch plane at BECO. The most significant high-frequency vibration occurred at shaped charge firings (insulation panel jettison and *Atlas/Centaur* separation) and were comparable to previous flights.

The *Surveyor* compartment pressure dropped from atmospheric to essentially zero within the first 115 sec of flight. As expected, there was no indication of pressure buildup in the spacecraft cavity at nose-fairing thruster bottle actuation.

6. Separation and Retromaneuver Systems

All vehicle separation systems, including shaped charge firing, functioned normally. Booster section jettison occurred as planned under sustainer engine pitch and yaw attitude control.

All four insulation panels are concluded to have separated and jettisoned normally, although 3 of 24 special break wires, installed for monitoring deployment, indicated abnormal separation motion. These abnormalities are attributed to damaged instrumentation due to debris ejected from shaped charge firing. The uniform compatibility of the 21 breakwires indicating normal panel jettison contradicts and invalidates the data from the remaining breakwire instrumentation.

Nose fairing separation occurred with a greater transient in vehicle roll than experienced on previous flights, but this was well within the *Atlas* control capability and produced no detrimental effects.

Atlas/Centaur separation occurred as planned, with all eight *Atlas* retrorockets firing. *Atlas* motion in the critical pitch plane was the smallest yet observed during staging.

At spacecraft separation, all three pyrotechnic release latches actuated within 2 millisec of each other. The spring assemblies extended normally, producing a spacecraft separation rate of about 1 ft/sec. Residual *Centaur* rotation at separation was 0.16 deg/sec. Spring stroke vs time was nearly identical for the three springs, resulting in a net total spacecraft rotational rate of 0.34 deg/sec, calculated using linear potentiometer data. Spacecraft gyro data confirmed this low angular rate, which compares very favorably with the specified maximum rate of 3.0 deg/sec.

The *Centaur* retromaneuver was executed as planned. Five hours after spacecraft separation the distance between the *Centaur* and spacecraft had increased to 1054 km, which compares favorably with an expected nominal separation of 960 km and a required separation of 336 km at that time. Closest approach of the *Centaur* to the moon occurred at 12:37:06 GMT on June 2, 1966, with a miss distance of 16,945 km (see Fig. VII-5).

IV. SURVEYOR SPACECRAFT

A. Spacecraft System

In the *Surveyor* design, the primary objective was to maximize the probability of successful spacecraft operation within the basic limitations imposed by launch vehicle capabilities, the extent of knowledge of transit and lunar environments, and the current technological state of the art. In keeping with this primary objective, design policies were established which (1) minimized spacecraft complexity by placing responsibility for mission control and decision making on earth-based equipment wherever possible, (2) provided the capability of transmitting a large number of different data channels from the spacecraft, (3) included provisions for accommodating a large number of individual commands from the earth, and (4) made all subsystems as autonomous as practicable.

The peripheral areas shown in Fig. IV-I represent all of the subsystems essential to the operation of the *Surveyor* spacecraft. These areas are discussed in subsequent sections. The *Surveyor I* spacecraft is an engineering test model equipped with all subsystems necessary for a soft-landing on the moon. It carries a survey television system and engineering instrumentation, including temperature sensors, strain gages, accelerometers, and position-indicating devices. The spacecraft coordinate system is an orthogonal, right-hand Cartesian coordinate

system. The physical configuration of the spacecraft is shown in Fig. IV-2 and in Appendix B. A simplified functional block diagram of the spacecraft system is shown in Fig. IV-3.

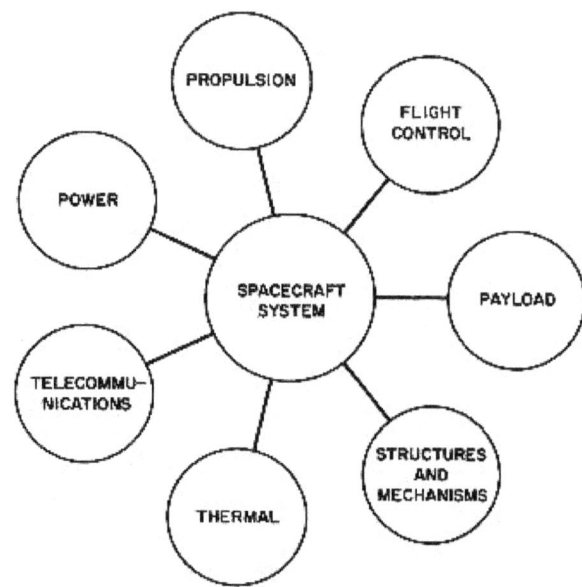

Fig. IV-1. Spacecraft subsystems

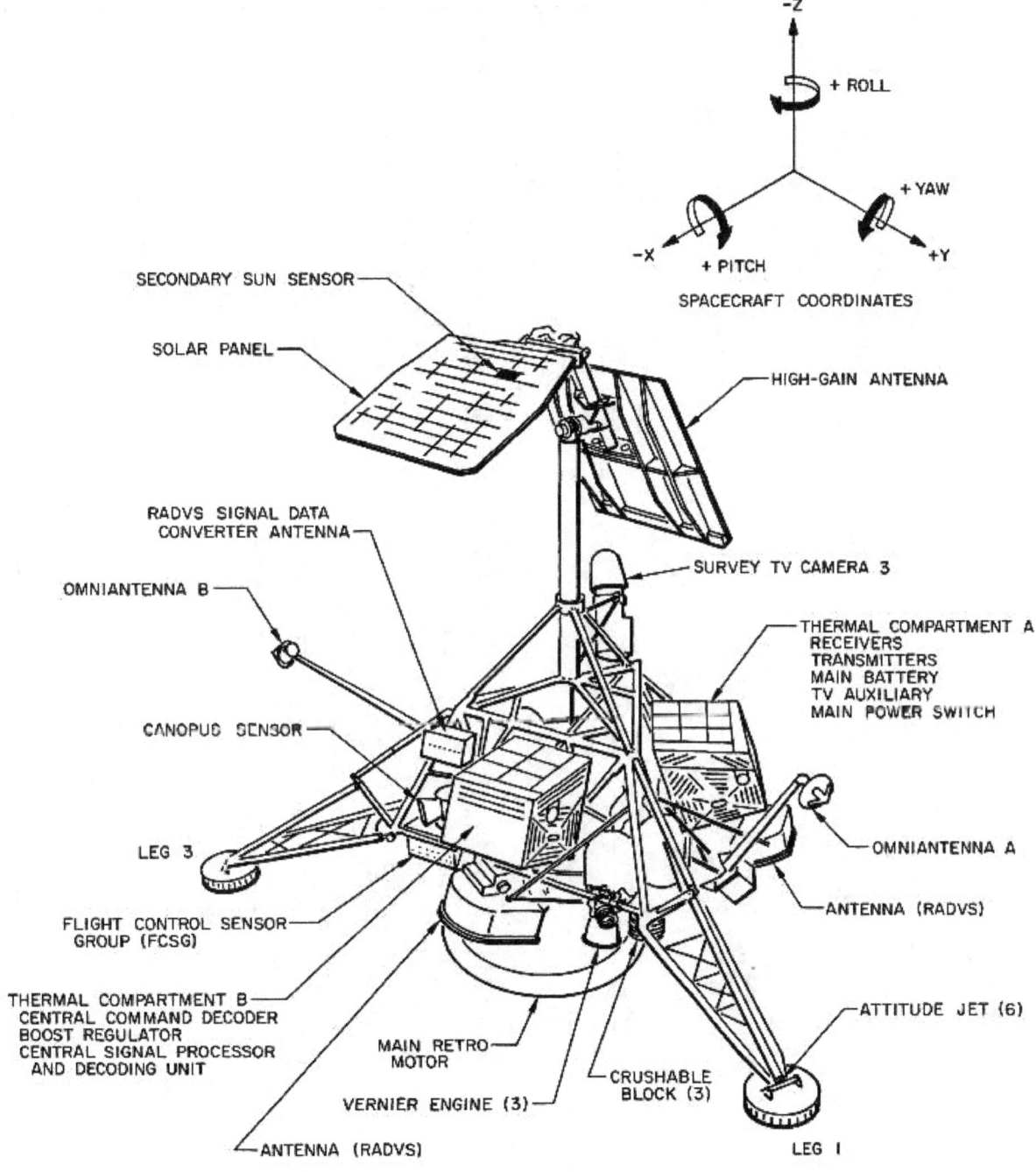

-Z

+ ROLL

+ YAW

-X

+ PITCH

+Y

SPACECRAFT COORDINATES

SECONDARY SUN SENSOR

SOLAR PANEL

HIGH-GAIN ANTENNA

RADVS SIGNAL DATA
CONVERTER ANTENNA

SURVEY TV CAMERA 3

OMNIANTENNA B

THERMAL COMPARTMENT A
RECEIVERS
TRANSMITTERS
MAIN BATTERY
TV AUXILIARY
MAIN POWER SWITCH

CANOPUS SENSOR

LEG 3

OMNIANTENNA A

ANTENNA (RADVS)

FLIGHT CONTROL SENSOR
GROUP (FCSG)

THERMAL COMPARTMENT B
CENTRAL COMMAND DECODER
BOOST REGULATOR
CENTRAL SIGNAL PROCESSOR
AND DECODING UNIT

ATTITUDE JET (6)

MAIN RETRO
MOTOR

CRUSHABLE
BLOCK (3)

VERNIER ENGINE (3)

ANTENNA (RADVS)

LEG I

Fig. IV-2. *Surveyor I spacecraft configuration*

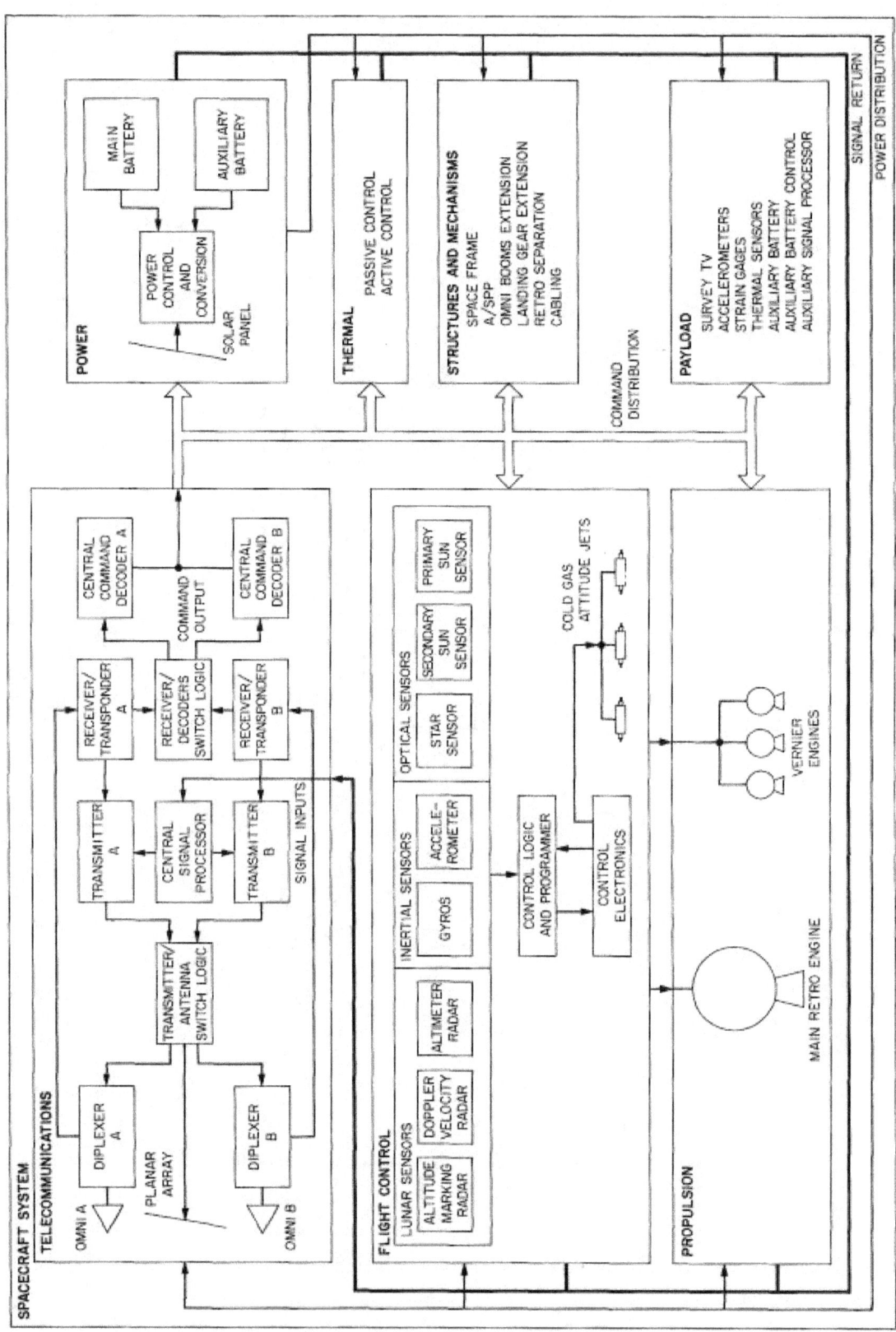

Fig. IV-3. Simplified spacecraft functional block diagram

1. Spacecraft Mass Properties

Spacecraft center-of-gravity and moment-of-inertia conditions are discussed for three major configurations: (1) stowed for launch at liftoff weight, (2) deployed for midcourse and retro maneuvers at separated weight, and (3) deployed for touchdown at landed weight. The limits of the center-of-gravity travel in the X-Y plane are defined by the first two conditions, and the limits of center-of-gravity travel along the Z-axis are defined by the first and last conditions. The center-of-gravity envelope has a radius of 1 in. and a length of 3.5 in. (16 in. to 19.5 in. above the centerline of the plane containing the spacecraft leg hinge-points). Center-of-gravity limits after *Surveyor/Centaur* separation for midcourse and retro maneuvers are limited by the attitude correction capabilities of the flight control and vernier engine subsystems during retrorocket burning. The limits of travel of the vertical center-of-gravity location in the touchdown configuration are designed to the landing site assumptions, so that the spacecraft will not topple when landing.

Flight performance during the terminal sequence indicated both sufficient accuracy of the CG determination process and the ability of the system to readily handle minor anomalies. The CG offset caused by the stowed Antenna A was calculated to be about 0.030 in. during midcourse and 0.115 in. after retro ejection, both well within the tolerance circle. In addition to verifying the alignment integrity of the spacecraft design, the stable terminal sequence performance also verified the assumption that the retrorocket thrust vector does coincide with the exit cone center line.

Surveyor I weighed 2,192.86 lb at launch. For the stowed configuration (landing gear and omniantennas up, antenna/solar panel positioner stowed) the moments of inertia about the center of gravity are $I_{xx} = 182.9$ slug-ft^2, $I_{yy} = 180.8$ slug-ft^2, $I_{zz} = 188.3$ slug-ft^2. For the transit configuration (landing gear and omniantennas down, A/SPP deployed), $I_{xx} = 214.9$ slug-ft^2, $I_{yy} = 200.2$ slug-ft^2, $I_{zz} = 218.4$ slug-ft^2.

2. Spacecraft Power Management

The prime source of electrical power for the spacecraft during transit is the solar panel. A main battery provides power to the 22-v unregulated bus to augment or replace the solar panel output. It is charged by power from the solar panel in excess of the needs of the spacecraft. Another battery, the auxiliary battery, provides the additional peak load and power storage capability for the transit and landing phase. This battery is not rechargeable. Power is also available as regulated 29-v. The batteries provide about 4090 w (about 34%) of the spacecraft energy during transit, while the remainder is supplied by the solar panel.

The total remaining battery energy at touchdown was 74 ±15 amp-hr. The nominal prediction was 70.5 amp-hr. The solar panel output power at maximum output power point was approximately 90 w. The average OCR output power was 69.5 w, with an average efficiency of 78.5% for coast phase loads.

3. Spacecraft Thermal Control

Thermal control of the equipment over the extreme temperature range of the lunar surface (+260 to −260°F) is accomplished by a combination of passive, semipassive, and active methods. Because of the orientation of the spacecraft on the lunar surface, the sun traveled across the spacecraft directly along the Y-Y axis. This condition kept the TV survey camera and the compartments, especially Compartment A, well shaded by the solar panel and planar array antenna during a large part of the lunar day. Consequently, the spacecraft was operational for most of the lunar day. Post-touchdown thermal orientation of the spacecraft is shown graphically in Fig. IV-4.

4. Spacecraft Instrumentation

Engineering instrumentation is included on the basic bus to monitor the performance of the spacecraft and its response to the environment. The engineering instrumentation consists of the following:

- 64 resistance thermal sensors
- 12 mechanical switches
- 12 current shunts
- 7 position potentiometers
- 4 pressure sensor potentiometers
- 4 accelerometers (with one accelerometer amplifier)
- 3 spin-motor running detectors
- 1 microdiode thermal sensor

5. Spacecraft Tracking

Tracking of the spacecraft is accomplished by doppler tracking. A transponder, operating in conjunction with a receiver and a transmitter, provides a phase-coherent system for doppler tracking of the spacecraft during transit. There are two complete receiver-transponder-transmitter systems to enhance reliability. (Fig. IV-3.)

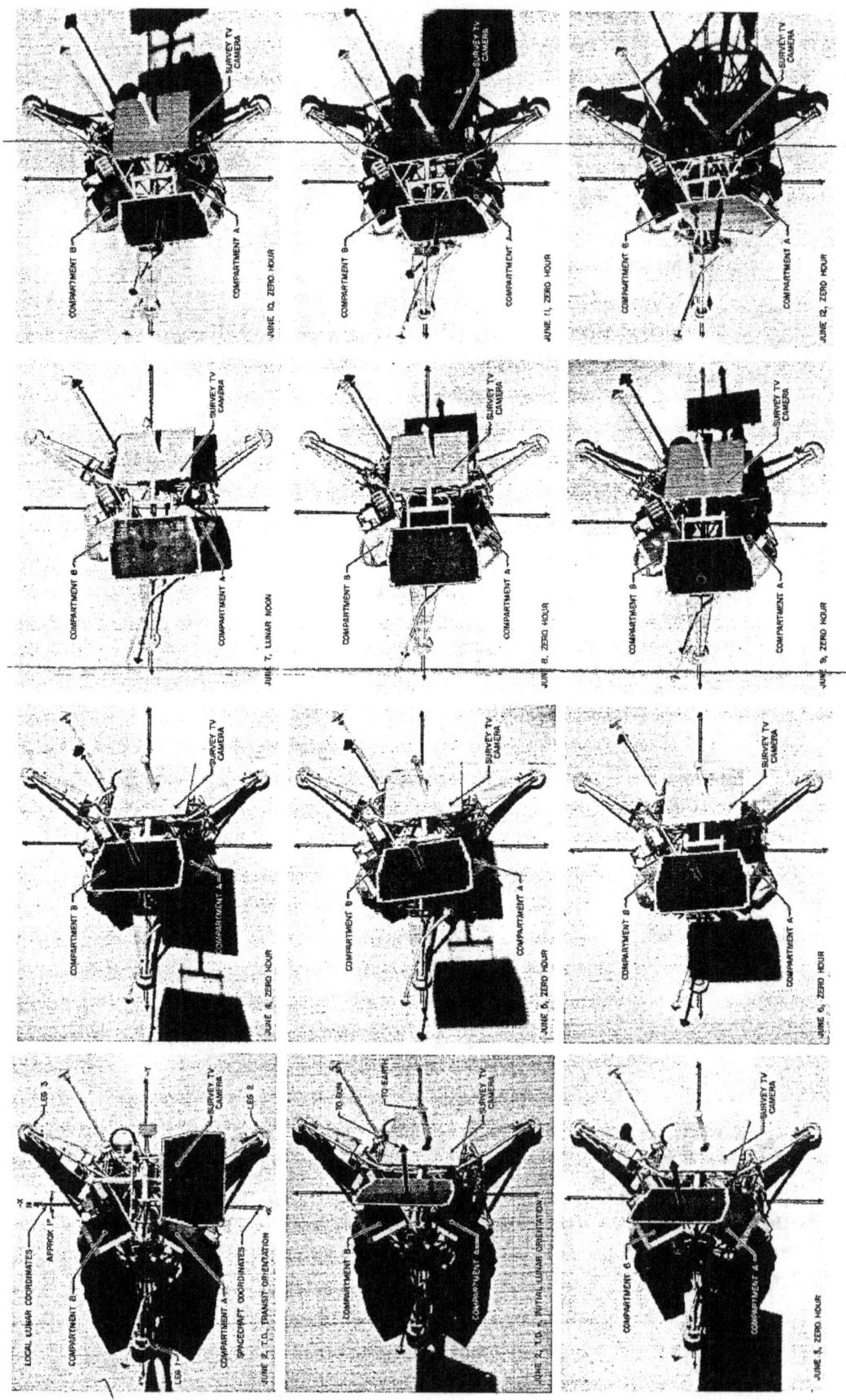

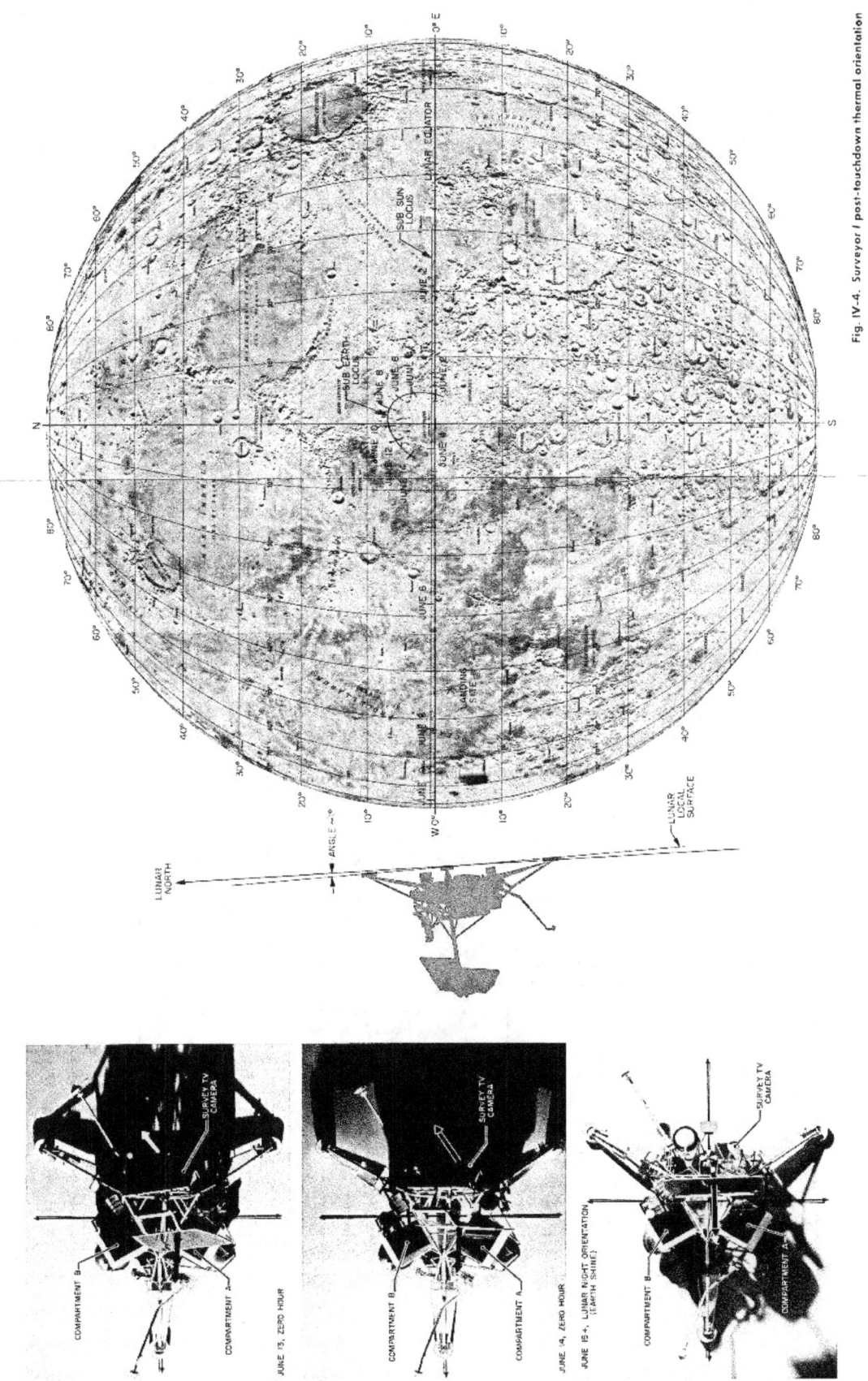

Fig. IV-4. Surveyor I post-touchdown thermal orientation

6. Spacecraft Communication

Communication with the spacecraft is accomplished by the RF link. Critical operations such as the midcourse maneuver and the descent phases are timed to coincide with control by the DSIF station at Goldstone. Two identical receivers (A and B) on board the spacecraft receive commands from ground control. Both receivers operate continuously throughout the life of the spacecraft to provide continuous command capability. The spacecraft processes data produced by the voltage, current, temperature, and pressure sensors, as well as the tele-

vision camera, accelerometers, and strain gages. This data is transmitted to ground stations by two identical transmitters (A and B) on board. The transmitter power available at the antenna varies from the nominal 100 mw (low power) and 10 w (high power) at the transmitter output.

The number of telemetry data channels operating simultaneously during the mission is controlled by ground commands. The channels required depend on the phase of the mission. There are seven commutators, as shown in Table IV-1. The commutators can be operated, upon

Table IV-1. Content of telemetry signals from spacecraft

Data	Source	Significance	Number of points sampled	Form	Comments
Commutator, Mode 1	Flight control, propulsion	Provides data required for midcourse maneuver	100	Digital	Modes 1, 2, 3, and 4 used one at a time on command per Standard Sequence of Events (SSE)
Commutator, Mode 2	Flight control, propulsion, approach TV, AMR, RADVS	Provides data required for retro descent	100	Digital	
Commutator, Mode 3	Inertial guidance, approach TV, AMR, RADVS, vernier engines	Provides data required for vernier descent	50	Digital	
Commutator, Mode 4	Temperatures, power status, telecommunications	Provides data required for miscellaneous transit and lunar surface operations	100	Digital	
Cruise phase commutator, Mode 5	Flight control, power status, temperature	Provides data required during cruise mode to determine general spacecraft status	120	Digital	Used on command per SSE
Thrust phase commutator, Mode 6	Flight control, power status, AMR, RADVS, vernier engine conditions	Provides data required for backup of Modes 1, 2, and 3 during thrusting maneuvers	120	Digital	Used on command per SSE
TV commutator, Mode 7	TV survey camera	Provides frame identification while survey TV is operating	16	Digital	Frame ID alternates with analog video signals
Vibration	Accelerometers	Indicates vibrations due to booster engine, main retro-rocket, and vernier engine firings and mechanical shock during landing	8	Analog	
Shock absorber data	Strain gages	Measures strain on landing gear due to landing shock	3	Digital	
Gyro speed	Inertial guidance unit	Indicates angular rate of gyro spin motors	3	Analog	Samples are pitch, roll, and yaw axes on command per SSE

earth command, at any of the following bit rates: 17 3/16, 137 1/2, 550, 1100, and 4400 bit/sec. The choice of the rate is determined on the basis of the strength of the received signal at the DSIF station.

7. Spacecraft Attitude Control

Attitude control of the spacecraft is accomplished by the flight control function. The spacecraft detects attitude by celestial, inertial, and radar sensors. The sensor outputs are processed to create thrust commands to orient and stabilize the spacecraft during transit and to provide trajectory correction and establish vernier descent control for soft-landing the spacecraft on the lunar surface. Fig. IV-5 shows the spacecraft motion about its coordinate axes relative to the celestial references. The cone angle of the earth is the angle between the sun vector and the earth vector as seen from the spacecraft. The clock angle of the earth is measured in a plane perpendicular to the sun vector from the projection of the star Canopus vector to the projection of the earth vector in the plane. The spacecraft coordinate system may be related to the cone and clock angles coordinate system, provided sun and Canopus lock-on has been achieved. Note that the spacecraft minus Z-axis is directed toward the sun and the minus X-axis is coincident with the

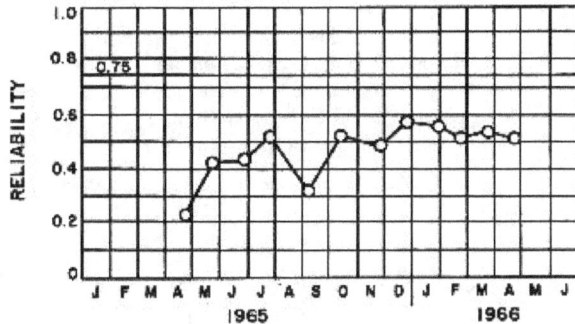

Fig. IV-6. *Surveyor I* spacecraft reliability test history

projection of the Canopus vector in the plane perpendicular to the direction of the sun.

8. Spacecraft Reliability

The prelaunch reliability estimate for *Surveyor I* was 0.51 for the flight and landing mission. The estimate is based on systems test data. Owing to the number of unit changes on the spacecraft, the reliability estimate is considered generic to *Surveyor I* rather than descriptive of the exact *Surveyor I* spacecraft configuration. Figure IV-6 shows the history of reliability estimates for *Surveyor I* during its system test phases.

The detail reliability estimates for flight and landing are listed in Table IV-2.

Table IV-2. *Surveyor I* spacecraft reliability (flight and landing)

Subsystem	Reliability estimates
Telecommunications	0.922
Vehicle mechanisms	0.854
Propulsion	0.991
Electrical power	0.866
Flight controls	0.954
Spacecraft	0.645
(System interaction reliability factor)	(0.788)
Spacecraft reliability = (0.645) (0.788) = 0.51	

9. Spacecraft Initial Launch Conditions

At the time of launch, the spacecraft main battery is fully charged and the Canopus sensor field of view is adjusted for the particular launch window. The temperatures of the main retro engine, vernier engine propellant, shock absorbers, Compartments A and B, flight control electronics, Canopus sensor, gyros, and the auxiliary bat-

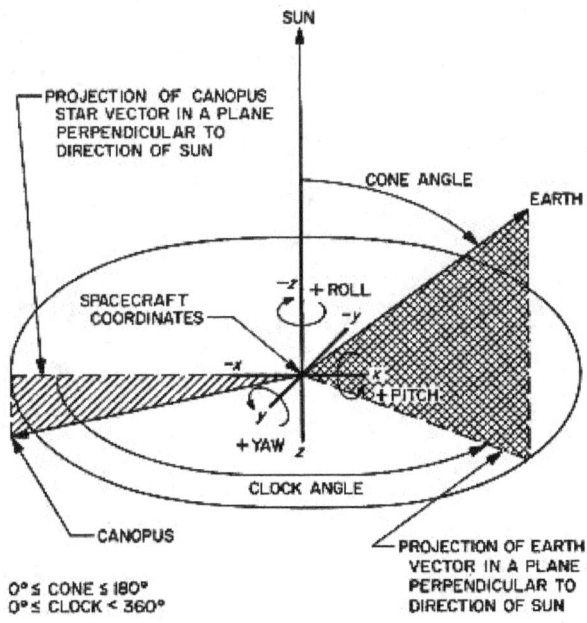

Fig. IV-5. Spacecraft coordinates relative to celestial references

tery are checked. Once the spacecraft is launched, external control is not available until the initial DSIF acquisition is accomplished. Because of these requirements, the following spacecraft operating conditions are established during the countdown:

1. Flight control is *on* and in the rate-stabilized mode to null out angular rates imparted to the spacecraft at separation from the *Centaur*.

2. The attitude jets are inhibited during boost.

3. Accelerometer amplifiers are *on* to permit accelerometer data to be transmitted through the *Centaur* telemetry channels.

4. One transmitter, receiver, and transponder are *on* to facilitate initial DSIF acquisition.

5. The transmitter traveling wave tube (TWT) filament is *on* to permit high-power operation when the *Centaur* preseparation command for high power is given.

6. The spacecraft signal processor is in the *coast phase* mode to provide pulse code modulation (PCM) data

through the *Centaur* and the spacecraft telemetry links during the period between boost and DSIF acquisition.

10. Mission Summary

Following a successful prelaunch Systems Readiness Test (SRT) and countdown, the *Surveyor* spacecraft was launched into a nearly perfect lunar trajectory by an *Atlas/Centaur* (A/C-10) vehicle at 1441 GMT (0741 PDT) on May 30, 1966. Sixty-three hours and thirty-six minutes later, after a nearly ideal transit phase, marred only by a failure in the Omniantenna A mechanism, *Surveyor I* landed softly on the lunar surface within a few kilometers of the prescribed landing spot and without even a momentary loss in the transmission of telemetry data from spacecraft to earth. A typical transit sequence is shown in Fig. IV-7.

11. Spacecraft Anomalies

The only significant spacecraft anomaly in the whole transit mission was the failure of the Omniantenna A

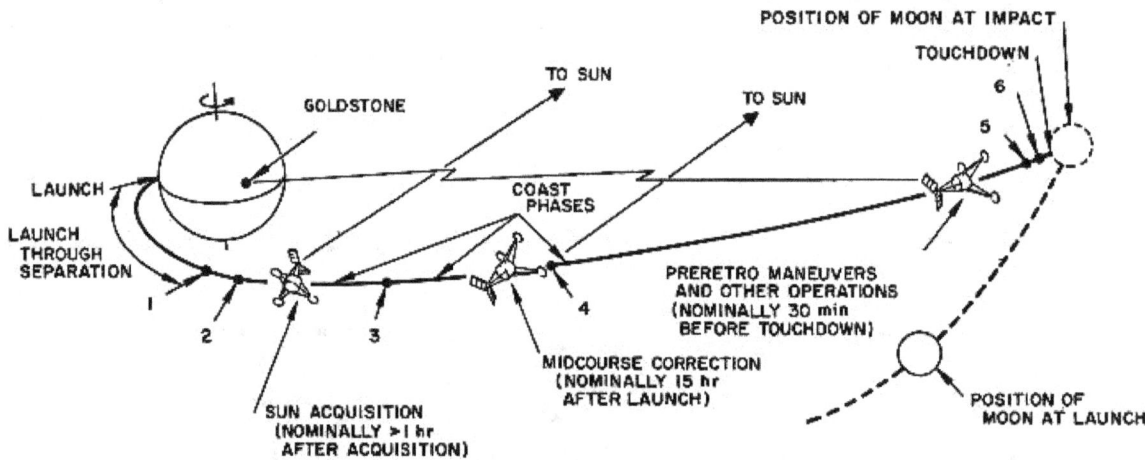

I INJECTION AND SEPARATION
2 INITIAL DSIF ACQUISITION
3 STAR ACQUISITION AND VERIFICATION
  (NOMINALLY 15 hr AFTER LAUNCH)
4 REACQUISITION OF SUN AND STAR
  (IMMEDIATELY AFTER MIDCOURSE CORRECTION)
5 RETRO PHASE (INITIATED AT NOMINAL RANGE
  OF 60 mi FROM MOON)
6 VERNIER DESCENT (NOMINALLY LAST 35,000 ft
  OF FLIGHT)

Fig. IV-7. Typical transit sequence

mechanism. This failure gave very little trouble throughout the mission as all objectives were able to be met using terminal maneuvers were selected which resulted in a vehicle roll orientation on the moon which precluded the possibility of obtaining a TV picture of the earth. The significant changes in procedure from the prepared Omniantenna B. Because Omniantenna A failed to deploy, standard mission sequence were:

1. Failure of Omniantenna A to extend during the pre-separation phase, thus causing the implementation of nonstandard sequences to attempt to extend the omniantenna boom.

2. Performance of all high-power interrogations (except the one conducted as part of the terminal descent sequence) in low power because of the good telecommunication signal-strength operating margin.

3. Use of manual lock-on (instead of automatic lock-on) for Canopus acquisition owing to high signal level produced by all stars, including Canopus.

4. Additional gyro drift checks obtained during Coast Phase II to obtain the best estimate of the drift of each gyro for use in the terminal descent computations.

5. Additional use of auxiliary battery mode (instead of high current mode) to increase the rate of temperature rise of the auxiliary battery so that the desired auxiliary battery temperature was achieved during the terminal descent.

6. Three attitude maneuvers performed during terminal descent because of the constraint resulting from not using Omniantenna A.

7. Post-touchdown turn-on of the approach camera system to verify that it had survived.

12. Spacecraft Performance

All spacecraft performance requirements were satisfactorily met. Table IV-3 summarizes the system performance requirements and the actual flight performance values determined from preliminary data reduction.

Following is a summary of the Surveyor I spacecraft performance as related to major flight events.

a. Boost phase. During the launch phase of the transit, telemetering of data is accomplished via the Centaur. Shock and vibration data are transmitted during this time.

The Surveyor spacecraft was subjected to a variable vibration environment during the boost phase which

Table IV-3. Surveyor I overall system performance summary

Item	Required	Actual
Injected weight	≤ 2250 lb	2192.86 lb
Mission payload weight	≥ 62.5 lb	64.10 lb
Spacecraft CG to roll axis offset	< 1 in.	0.099 in. at launch 0.148 in. during coast
Transit time	66 hr nominal	63 hr, 36 min
Landing accuracy	< 100 km, 3-σ for 50 m/sec M/C tracking errors)	
Lunar approach angle	0 to 25 deg	
Touchdown conditions		
Vertical velocity	< 15 ft/sec	12 ft/sec
Lateral velocity	< 5 ft/sec	2 ft/sec
Vertical shock (rigid body at CG)	< 30 Earth g	6.5 g (est.)
Lateral shock (rigid body at CG)	< 12 Earth g	1.5 g (est.)
Stable	Yes	Yes
Lunar operations	Sunrise to sunset	Yes
Lunar operations	150 hr of post-sunset	Yes

consisted of acoustically induced random vibration and the transient response to discrete flight events. It was instrumented with accelerometers in order to monitor this environment. However, the accelerometers were in the Z-direction only, and lateral axis vibration was not monitored.

The spacecraft Z-axis random and sine vibration flight environments at the spacecraft/adapter interface (Figs. IV-8 and IV-9) were within the design levels specified and within the 95% environment as estimated from Surveyor data using past Ranger and Surveyor dynamic model vibration data for statistical parameters. The spacecraft Z-axis shock spectra at the spacecraft/adapter interface exceeded the design shock spectra level by a large factor in the 600-cps region (Fig. IV-10).

b. Midcourse maneuver phase. All midcourse operations were performed normally. With the spacecraft visible to and controlled by DSS-11 (Goldstone, Calif.), the maneuver sequence for applying the midcourse thrust in the proper direction was initiated, with a minus roll of 86.5 deg and a minus yaw of 57.9 deg. With the vehicle

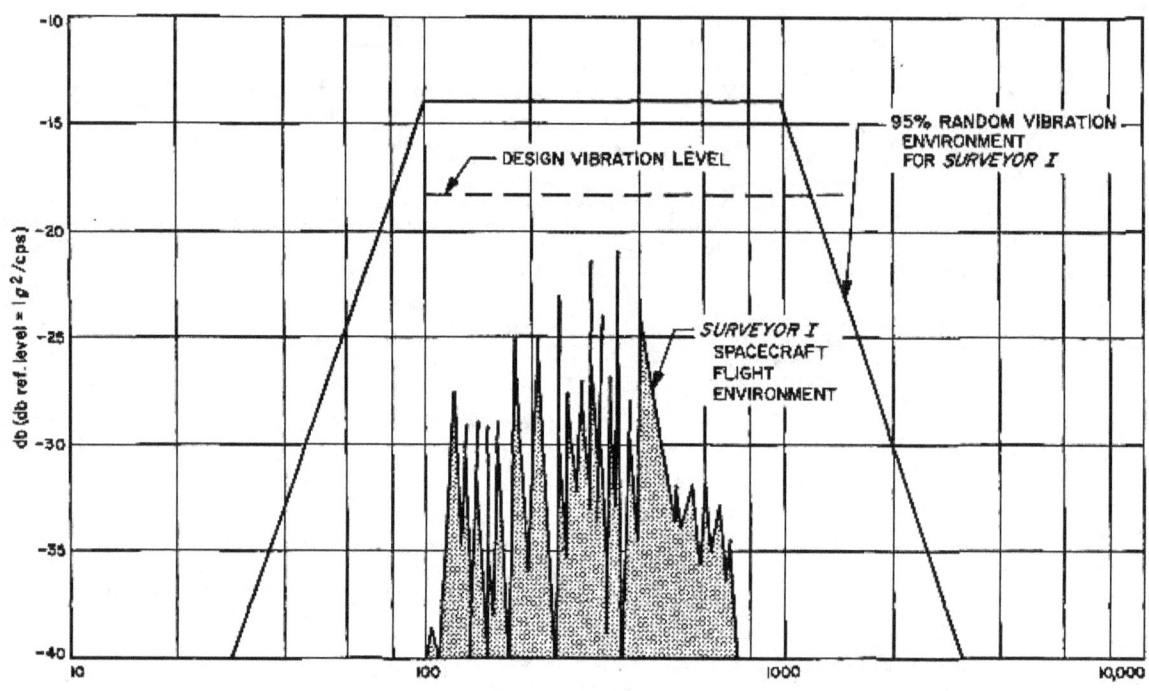

Fig. IV-8. Z-axis random environment at spacecraft/adapter interface

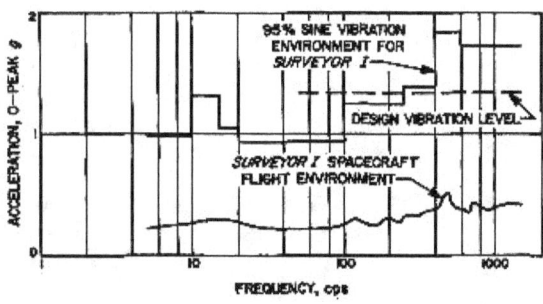

Fig. IV-9. Z-axis sine environment at spacecraft/adapter interface

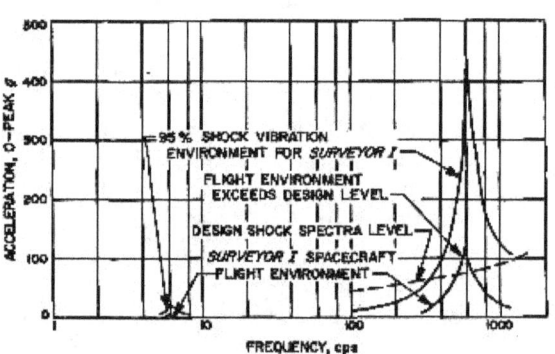

Fig. IV-10. Z-axis shock spectra environment at spacecraft/adapter interface

thrusting direction positioned properly, the midcourse thrust was applied by igniting the vernier engines and controlling their thrust to achieve a constant spacecraft acceleration of 0.1 g for 20.8 sec (a correction of 20.34 m/sec). The spacecraft was then repositioned for transit orientation. The lunar landing point errors from all sources were estimated to be 14 km -3σ, compared to a premaneuver estimate of 78 km -3σ. Tables IV-4 and

IV-5 summarize the predicted and actual midcourse thrust vector angle and magnitude errors contributed by various elements of the spacecraft system.

c. *Terminal descent phase.* The automatic sequence of events (including vernier ignition, retro ignition, RADVS,

Table IV-4. Major midcourse thrust vector
pointing angle errors

Item	Pointing angle errors, deg	
	3σ	Actual (best est. to date)
Retro engine		
Alignment to spacecraft	0.18	0.01
Flight control sensor group		
Alignment to spacecraft	0.20	0.03
Null offset	0.10	0.01
Limit cycle	0.18	0.09
Gyro drift	0.25	0.05
Flight control circuit accuracy	0.10	0.03
Total, deg (RSS)	$\sqrt{2} \times 0.43 = 0.60$ (weighted)	0.11
Miss on lunar surface, km	68.0	12.2

Table IV-5. Major midcourse thrust vector
magnitude errors

Item	Magnitude error, ft/sec	
	3σ	Actual
Velocity increment	164 (50 m/sec)	66.7 (20.35 m/sec)
Thrust offset	0.41	0.02
Accelerometer Misalignment		
Amplifier error	1.85	0.15
Reference voltage		
Scale factor		
Timing error	0.43	0.05
Engine start/stop transient	0.25	0.05
Total, ft/sec (RSS)	1.91	0.17
Miss on lunar surface, km		6.5

main retro burnout and separation, enable doppler control, 1000-ft mark, 10-ft/sec mark, and, finally, the 14-ft mark) occurred normally.

The vernier and main retro ignitions were relatively smooth, with only very small transient moment disturbances. External disturbances throughout the main retro phase were well within the capability of the control system, and were about 1 ft-lb in roll (primarily due to predictable engine bracket bending) and virtually none in pitch and yaw (indicating excellent prelaunch alignment of the retromotor and only a very small CG shift during the retro phase). A maximum roll actuator angle of 1.3 deg occurred during the main retro phase. With this exception, the actuator angle was less than 1.1 deg throughout the mission.

Table IV-6 summarizes the predicted main retro burnout dispersion errors contributed by various error sources and the estimated allocation to these sources of flight errors determined from the known spacecraft velocity and altitude at the time of main retro burnout.

d. Touchdown phase. Touchdown occurred between 1 and 2 sec following the occurrence of the 14-ft mark. The altitude of vernier engine cutoff was estimated at 12 ft instead of 14 ft. Confirmation of the soft landing was provided by the fact that the DSIF receiver retained its lock-on to the spacecraft transmitter signal and by the low magnitude of the force telemetered by the shock-absorber strain gages (< 1600 lb, indicating touchdown velocities on the order of 12 ft/sec). It was determined that *Surveyor I* landed on a slope approximately 1 deg from the local horizontal.

Table IV-6. Main retro burnout dispersions

Item	Main retro burnout value and dispersion							
	Nominal value	3σ	Velocity dispersion, ft/sec	Altitude dispersion, ft	Actual value	Error	Velocity dispersion, ft/sec	Altitude dispersion, ft
Unbraked impact velocity, ft/sec	8746.7	29	-29.0	1430	8710.6	36.1	-36.1	+1800
Ignition slant range, ft	248,350	1584	1.0	1638	247,900	250	+0.2	-260
Spacecraft weight, lb	—	2	13.5	325	—	2	-13.5	+325
Weight of retro, lb	1244.1	3.7	41.7	1000	1244.1	—	—	—
Vernier thrust, lb	196	20	42.8	807	194	2	+4	-80
Main retro I_{sp}, %	—	1.0	83.6	2259	—	-0.25	+21	-565
Burning time, sec	38.5	1.5	3.4	7200	39.2	+0.7	+1.6	-3300
Total Dispersion	—	—	108	2200	—	—	-23	-2080

e. Lunar phase. Compartment thermal performance was greatly enhanced by planar array/solar panel shadowing for a good part of the lunar day. Compartment A reached a minimum on June 4 (top tray 50°F, battery 64°F) and continued rising until June 12 (top tray 83°F, battery 92°F). Similarly, Compartment B reached a minimum on June 2 (top tray 70°F, bottom tray 76°F, boost regulator 82°F) and reached a maximum on June 10 (top tray 118°F, bottom tray 123°F, boost regulator 132°F).

Thrust chamber assembly (TCA) temperatures did not reach the predicted levels subsequent to lunar surface touchdown. TCA's were expected to reach peak temperatures of approximately 350°F on the lunar surface. However, the maximum temperature was observed to be 243°F. Maximum temperatures of the helium tank, propellant lines, and propellant tanks were below predictions.

The thermal behavior of the TV survey camera was such that it permitted unlimited intermittent operation on the lunar surface. Shadowing of the camera during the lunar day maintained camera temperatures below the upper operational limits. During the first lunar day, the TV survey camera system accepted a total of approximately 86,000 ground commands and transmitted about 10,000 TV frames.

Interrogation of the spacecraft was stopped at 20:30 GMT, June 16, 1966. The spacecraft survived lunar night and achieved operational capability on the second lunar day.

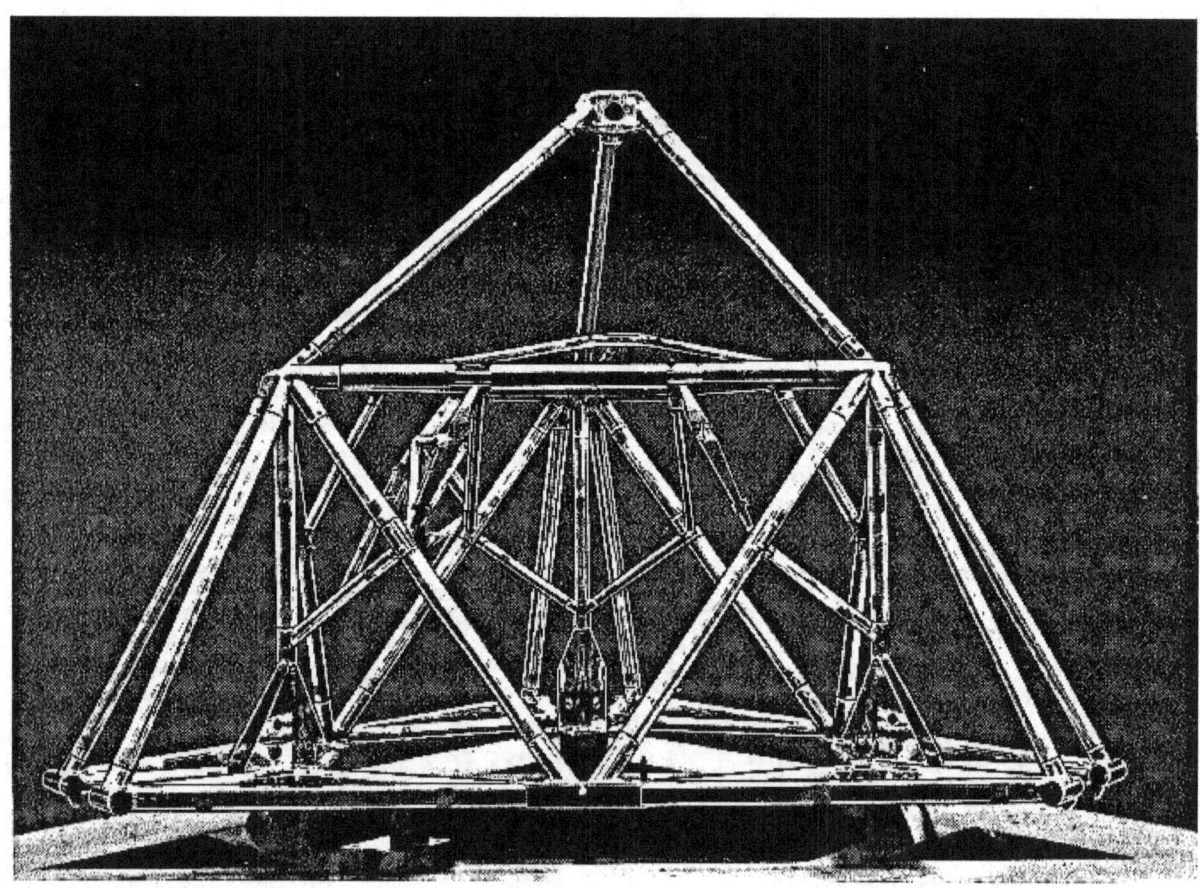

Fig. IV-11. Spaceframe

B. Structures and Mechanisms

The vehicle and mechanisms subsystem provides support, alignment, thermal protection, electrical interconnection, mechanical actuation, and touchdown stabilization for the spacecraft and its components. The subsystem includes the basic spaceframe, landing gear mechanism, crushable blocks, omnidirectional antenna mechanisms, antenna/solar panel positioner (A/SPP), pyrotechnic devices, electronic packaging and cabling, thermal compartments, thermal switches, separation sensing and arming device, and secondary sun sensor.

1. Spaceframe and Substructure

The spaceframe (Fig. IV-11), constructed of thin-wall aluminum tubing, is the basic structure of the spacecraft. The substructure is used to provide attachment between some subsystems and the spaceframe. The landing legs and crushable blocks, the retrorocket engine, the *Centaur* interconnect structure, the vernier propulsion engines and tanks, and the A/SPP attach directly to the spaceframe. The substructure is used for the thermal compartments, TV subsystem, auxiliary battery, RADVS antennas, flight control sensor group, attitude control nitrogen tank, and the vernier system helium tank.

Gyro data and linear potentiometer data indicated that the two separation events, spacecraft from the *Centaur* adapter and retromotor from the spacecraft, occurred as predicted without physical contact of one body with the other after initiation of separation.

The spacecraft survived the boost, cruise, midcourse and retro phase of the flight with no obvious anomalies. The vibration accelerations were significant only during the boost phase. Comparison of the flight versus the predicted flight environment can only be made in the Z-direction from the flight data. In the frequency range of 5-100 cps, the predicted flight environment comfortably envelops the measured environment over most of the frequency range (Figs. IV-8, -9).

In general, the performance of *Surveyor I* during the final descent and landing phases was very close to nominal. The three landing strain gages, mounted on the shock absorber columns of the three landing legs returned excellent analog force vs time traces (Fig. IV-12).* They indicated that the three landing legs contacted the

*Computer-simulated force-time data is shown in Fig. IV-13.

lunar surface in the following sequence and developed maximum forces of:

Leg 2, 0.000 ref . . . . 1570±80 lb

Leg 1, 0.005±0.001 sec after Leg 2 . . . . . . 1380±70

Leg 3, 0.019±0.001 sec after Leg 1 . . . . . . 1400±70

The maximum forces occurred approximately 120 msec after impact, followed by a rapid decrease. At approximately 0.5 sec after impact, all three channels show zero force, indicating that the spacecraft rebounded off the lunar surface. A second impact was registered approximately 1.1 to 1.2 sec after the first. Maximum forces developed in the second impact reached approximately the 400-lb level. Several TV pictures show block imprints in the lunar surface, confirming the occurrence of block contact as indicated.

Table IV-7 lists the actual and predicted landing performance data.

The actual velocity at vernier cutoff listed above has been determined by a linear least squares fit in the RADVS range data. This velocity and the time differences between vernier cutoff and initial contact have been used to establish the vertical landing velocity and spacecraft altitude at vernier cutoff.

The predicted maximum shock absorber force is based on a hard surface landing. This prediction constitutes an upper bound since the landing simulation program does not represent the flexibility of the space frame and superstructure which, by absorbing some of the energy dissipated at landing, tends to decrease the shock absorber forces.

Table IV-7. Surveyor I performance data

Item	Actual	Predicted
Horizontal landing velocity, ft/sec	< 2.8	< 5.0
Vertical landing velocity, ft/sec	11.7±0.4	12.6±2.5
Vertical velocity at vernier cutoff, ft/sec	4.3±0.1	5±1.5
Altitude at vernier cutoff, ft	11.0±0.6	13±4.5
Maximum shock absorber force, lb	1570±80	1765
Angle between roll axis and vertical at landing, deg	< 2.8	< 7.0
Total angular velocity at landing, deg/sec	< 1.0	< 3.2

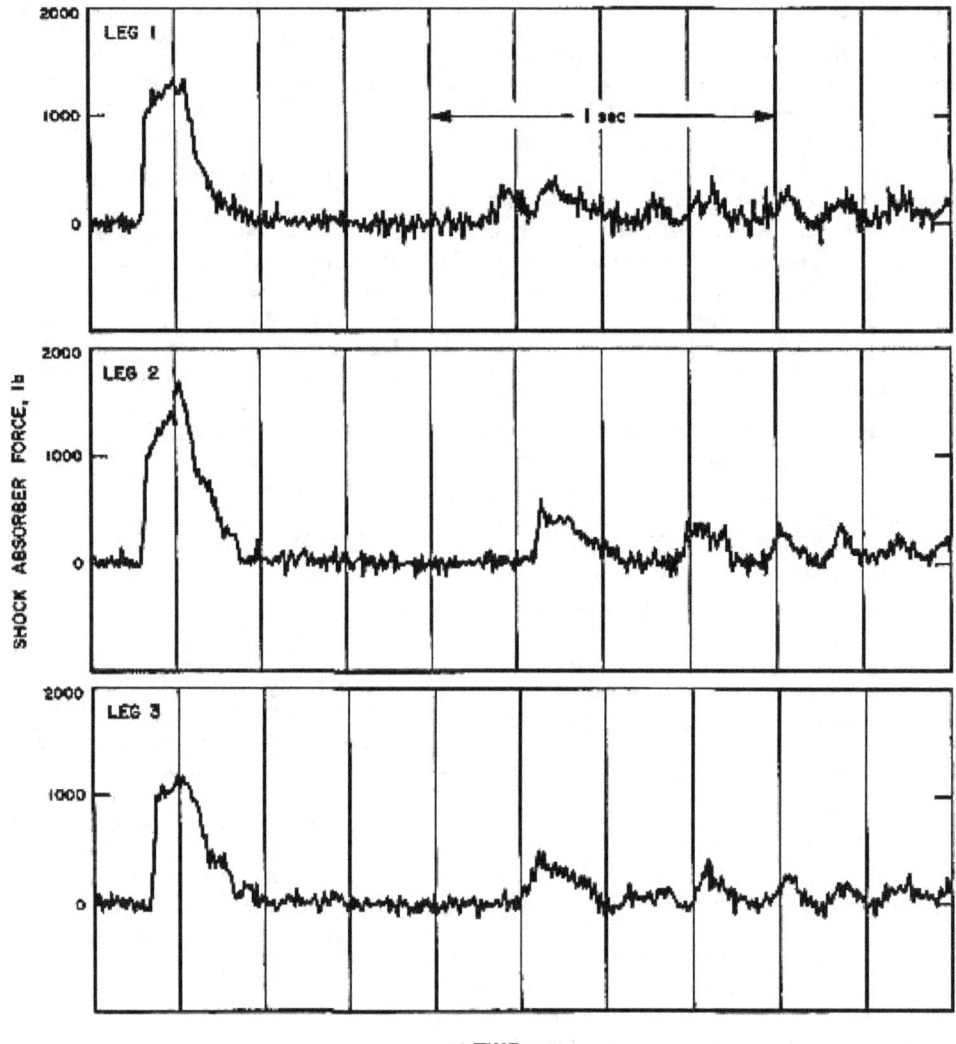

Fig. IV-12. Actual touchdown strain-gage data

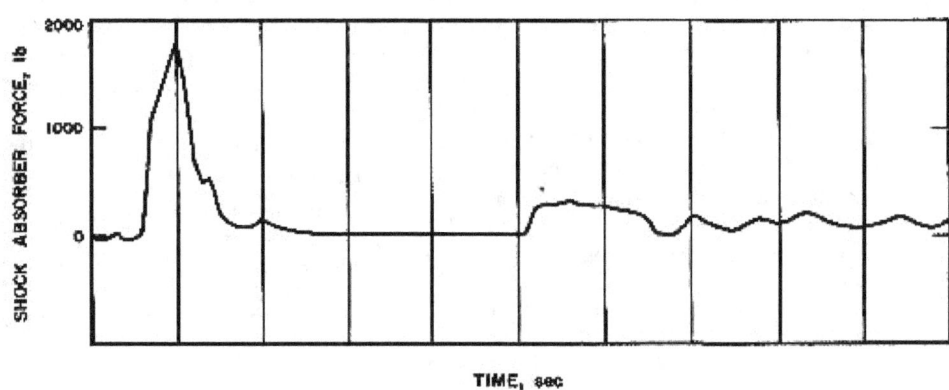

Fig. IV-13. Computer-simulated touchdown strain-gage data

2. Landing Gear and Crushable Blocks

The three landing-leg mechanisms are each made up of a landing leg, an intermediate A-frame, a shock absorber, a footpad, a locking strut, and a position potentiometer (Fig. IV-14).

The shock absorber, intermediate A-frame, and telescoping lock strut are interconnected to the spaceframe for folded stowage in the nose shroud. Torsion springs at the leg hinge extend the legs when the squib-actuated pin pullers are operated by *Centaur* or earth command. The hydraulic shock absorber compresses with landing load. The shock absorbers, foot pads, and crushable blocks are designed to absorb the landing shock. After landing, the shock absorbers are locked in place by squib-actuated pin locks.

Computer-simulated landings on a hard surface produce shock-absorber force-time histories (Fig. IV-13) very similar to that returned by *Surveyor I*. Shape and duration of the force pulse for first and second impact as well as the elapsed time between first impact, rebound, and second impact agree very closely, while the peak forces are approximately 20% higher in the hard-surface simulations. These forces depend to some degree on the behavior of the soil, but mainly on the landing gear design by which, under the conditions of the *Surveyor* landing, they were limited to approximately 500 lb per footpad, as indicated by landing simulations. Projected onto the frontal area of the footpads, this corresponds to a dynamic pressure of 8 to 10 psi.

3. Omnidirectional Antennas

The omnidirectional antennas are mounted on the ends of folding booms hinged to the spaceframe. Pins retain the booms in the stowed position. Squib-actuated pin pullers release the booms by *Centaur* or earth command, and torsion springs deploy the antennas. The booms are then locked in position.

The failure of Omniantenna A to deploy during transit represents the only anomaly detected for this mechanism. Several attempts to deploy Omniantenna A during transit failed. Shortly after touchdown, Omniantenna A indicated that it was extended. Omniantenna A fully deployed either at retro ignition or at touchdown, indicating thereby that the pin puller had fired previously. The probable explanation for this anomaly is that the boom of Omniantenna A was misaligned with the stowing bracket. Since the deployment spring produces only a 0.25-lb force, only a small misalignment is necessary to retain the boom.

Because Omniantenna A failed to deploy, the operational procedures during midcourse and terminal maneuvers were revised so as to obtain all required data

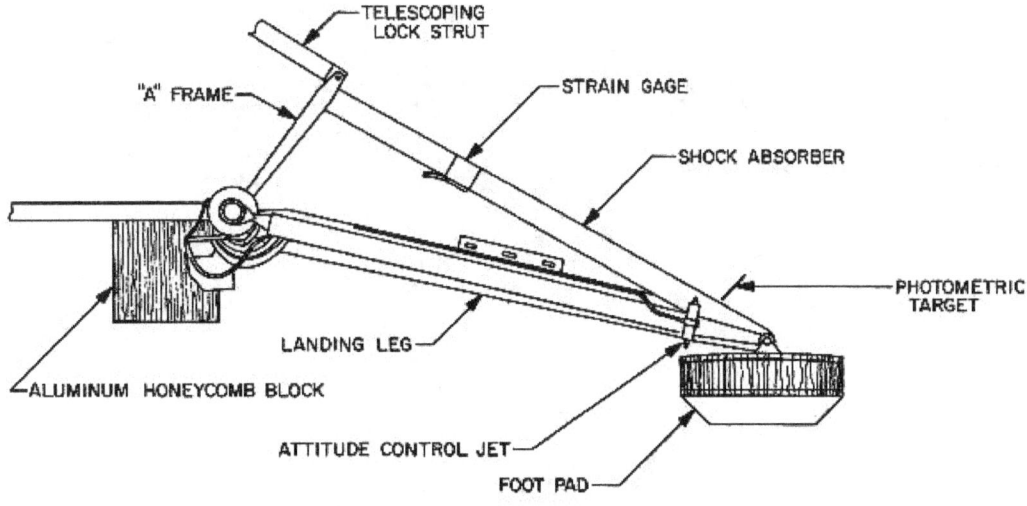

Fig. IV-14. Landing leg assembly

transmission from Omniantenna B. Maneuvers were selected to ensure good Omniantenna B coverage.

4. Antenna/Solar Panel Positioner (A/SPP)

The A/SPP supports and positions the high-gain planar array antenna and solar panel. The planar array antenna and solar panel have four axes of rotation: roll, polar, solar, and elevation (Fig. IV-15). Stepping motors rotate the axes in either direction in response to commands from earth or during automatic deployment following *Centaur/* spacecraft separation. This freedom of movement permits

orienting the planar array antenna toward the earth and the solar panel toward the sun.

The solar axis is locked in a vertical position for stowage in the nose shroud. After launch, the solar panel is positioned parallel to the spacecraft X axis. The A/SPP remains locked in this position until after touchdown, at which time the roll, solar, and elevation axes are released. Potentiometers on each axis are read to indicate A/SPP orientation. Each command from earth gives ⅛ degree of rotation in the roll, solar, and elevation axes and ¹⁄₁₆ degree in the polar axis.

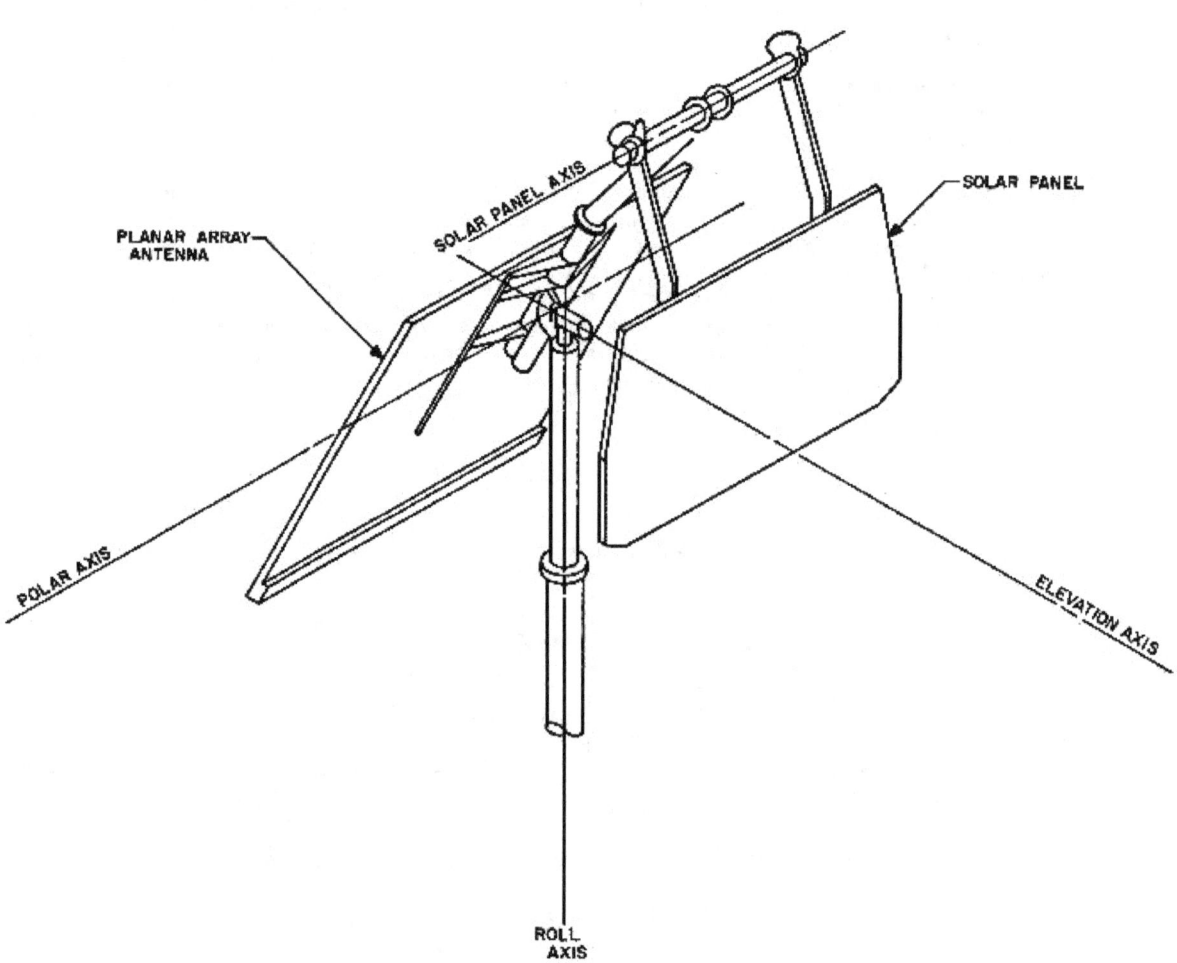

PLANAR ARRAY
ANTENNA

SOLAR PANEL AXIS

SOLAR PANEL

POLAR AXIS

ELEVATION AXIS

ROLL
AXIS

Fig. IV-15. Antenna/solar panel configuration

The A/SPP operated normally through the transit and post-touchdown operations. Evaluation of the data indicates continuous pulses of the stepping motors with few missed steps.

In addition to performing the regular functions that it was designed to perform, the A/SPP was used to perform two special tests. In the first test, the A/SPP rotated the solar panel to a position such that Compartment B was shadowed during lunar moon. The second test involved field-mapping of the planar array antenna to determine how closely it agreed with the standard antenna pattern obtained by averaging patterns from two other antennas.

At sundown, the solar panel was repositioned such that the light from the sun would strike it at a grazing angle just before noon. This was done to prevent a sizable solar panel output after lunar sunrise until the battery and electronics warmed up sufficiently to permit the battery to accept current from the solar panel without damage. The planar array was positioned to create the largest shadow during the morning of the beginning of the next lunar day. This will help another NASA project, *Lunar Orbiter*, in its attempt to photograph *Surveyor* as it passes over.

5. Thermal Compartments

Two thermal compartments (A and B) house thermally sensitive electronic items. Equipment in the compartments is mounted on thermal trays that distribute heat throughout the compartments. An insulating blanket, consisting of 75 sheets of 0.25-mil-thick aluminized mylar, is installed between the inner shell and the outer protective cover of the compartments. Compartment design employs thermal switches which are capable of varying the thermal conductance between the inner compartment and the external radiating surface. The thermal switches maintain thermal tray temperature below +125°F. Each compartment contains a thermal control and heater assembly to maintain the temperature of the thermal tray above a specified temperature (above 40°F for Compartment A and above 0°F for Compartment B). The thermal control and heater assembly is capable of automatic operation, or may be turned on or off by earth command. Components located within the compartments are identified in Table IV-8.

6. Thermal Switch

The thermal switch is a thermal-mechanical device which varies the conductive path between an external radiation surface and the top of the compartment (Fig.

Table IV-8. Thermal compartment component installation

Compartment A	Compartment B
Receivers (2)	Central command decoder
Transmitters (2)	Boost regulator
Main battery	Central signal processor
Battery charge regulator	Signal processing auxiliary
Engineering mechanisms auxiliary	Engineering signal processor
Television auxiliary	Low data rate auxiliary
Thermal control and heater assembly	Thermal control and heater assembly
Auxiliary battery control	Auxiliary engineering signal processor

IV-16). The switch is made up of two contact surfaces which are ground to within one wavelength of being optically flat. One surface is then coated with a conforming substance to form an intimate contact with the mating surface. The contact actuation is accomplished by four bimetallic elements located at the base of the switch. These elements are connected mechanically to the top of the compartment so that the compartment temperature controls the switch actuation. The switches are identical, but are adjusted to open at three different temperatures: 65, 50 and 40°F.

The external radiator surface is such that it absorbs only 12% of the solar energy incident on it and radiates 74% of the heat energy conducted to its surface. When the switch is closed and the compartment is hot, the switch loses its heat energy to space. When the compartment gets cold, the switch contacts open about 0.020 in., thereby opening the heat-conductive path to the radiator and thus reducing the heat loss through the switch to almost zero.

The thermal switches kept the electronics at or below the maximum temperature at all times during the lunar day, and therefore all switches were closed and had the proper conduction. However, a problem developed with the switches during the lunar night cycle. The switches were designed to open when the electronics module temperatures started to go below 65°F in Compartment A and below 50°F in Compartment B. Two switches in Compartment A did not open until the compartment reached 0°F. Two, or possibly three, of the Compartment B switches remained closed until after the last engineering interrogation, so it is not known at precisely what temperatures they opened, or whether they did open.

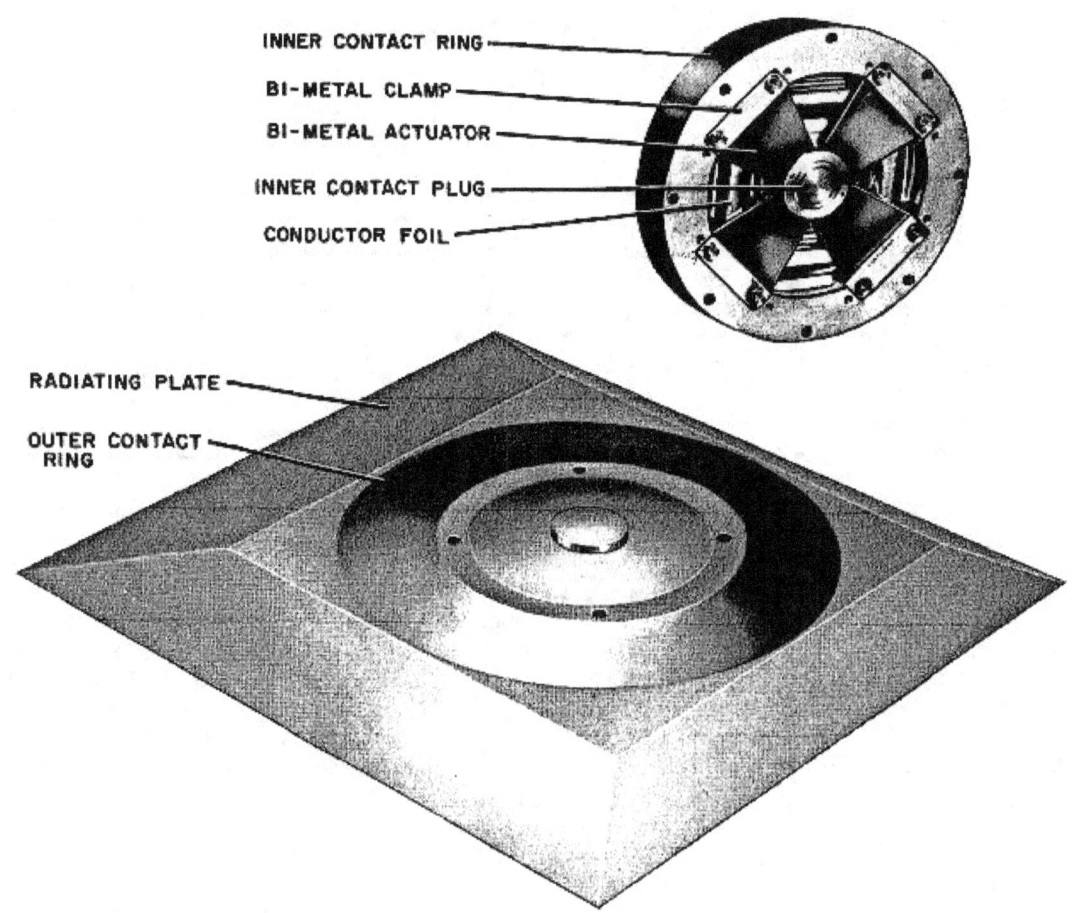

INNER CONTACT RING

BI-METAL CLAMP

BI-METAL ACTUATOR

INNER CONTACT PLUG

CONDUCTOR FOIL

RADIATING PLATE

OUTER CONTACT RING

Fig. IV-16. Thermal switch

7. Secondary Sun Sensor

The secondary sun sensor is an assembly of five cadmium-sulphide photoconductive cells mounted on the solar panel. On the lunar surface, the secondary sun sensor assists in determining spacecraft orientation with respect to the sun and aids in positioning the solar panel. Operation of the secondary sun sensor was nominal.

8. Pyrotechnic Devices

The pyrotechnic devices installed on *Surveyor I* are indicated in Table IV-9. All the squibs used in these devices are electrically initiated, hot-bridgewire, gas-generating devices. Qualification tests for flight squibs

included demonstration of reliability at a firing current level of 4 or 4.5 amp. "No Fire" tests were conducted at a 1-amp or 1-w level for 5 min. Electrical power required to initiate pyrotechnic devices is furnished by the spacecraft main battery. Power distribution is through 19.0- and 9.5-amp constant-current generators in the engineering mechanism auxiliary (EMA).

All pyrotechnic devices functioned normally upon command. Mechanical operation of locks, valves, switches, and plunger, actuated by squibs, was indicated on telemetry signals as part of the spacecraft engineering measurement data.

Table IV-9. Surveyor I pyrotechnic devices

Type	Location and use	Quantity of devices	Quantity of squibs	Command source
Pin pullers	Lock and release omnidirectional antennas A and B	2	2	Centaur programmer
Pin pullers	Lock and release landing legs	3	3	Centaur programmer
Pin pullers	Lock and release planar antenna and solar panel	7	7	Separation sensing and arming device and ground station
Pin puller	Lock and release vernier thrust chamber No. 1	1	1	Ground station
Separation nuts	Retro rocket attach and release	3	6	Flight control subsystem
Valve	Helium gas release and dump	1	2	Ground station
Pyro switches	EMA board No. 4, RADVS power on and off	4	4	Ground station and flight control subsystem
Initiator squibs	Safe and arm assembly retrorocket initiators	1	2	Flight control subsystem
Locking plungers	Landing leg, shock absorber locks	3	3	Ground station
		25	30	

9. Electronic Packaging and Cabling

The electronic assemblies for *Surveyor I* provided mechanical support for electronic components in order to insure proper operation throughout the various environmental conditions to which they were exposed during the mission. The assemblies (or control items) were constructed utilizing sheet metal structure, sandwich-type etched circuit board chassis with two-sided circuitry, plated through holes, and/or bifurcated terminals. Each control item, in general, consists of only a single functional subsystem and is located either in or out of the two thermally controlled compartments, depending on the temperature sensitivity of the particular subsystem. Electrical interconnection is accomplished primarily through the main spacecraft harness. The cabling system is constructed utilizing a light-weight, minimum-bulk, and abrasion-resistant wire which is an extruded teflon having a dip coating of modified polyimide.

C. Thermal

The thermal control subsystem is designed to provide acceptable thermal environments for all components during all phases of spacecraft operation. Spacecraft items with close temperature tolerances were grouped together in thermally controlled compartments. Those items with wide temperature tolerances were thermally decoupled from the compartments. The thermal design fits the "basic bus" concept in that the design was conceived to require minimum thermal design changes for future missions. Monitoring of the performance of the spacecraft thermal design is done by 74 temperature sensors which are distributed throughout the spacecraft as follows:

Flight control	6
Mechanisms	3
Radar	6
Electrical power	3
Transmitters	2
Approach TV	1
Survey TV	2
Vehicle structure	25
Propulsion	15
Engineering payload	11

The spacecraft thermal control subsystem is designed to function in the space environment, both in transit and on the lunar surface. Extremes in the environment as well as mission requirements on various pieces of the spacecraft have led to a variety of methods of thermal control. The spacecraft thermal control design is based upon the absorption, generation, conduction, and radiation of heat. Figure IV-17 shows those areas of the spacecraft serviced by different thermal designs.

The radiative properties of the external surfaces of major items are controlled by using paints, by polishing, and by using various other surface treatments. Reflecting mirrors are used to direct sunlight to certain components. In cases where the required radiative isolation cannot be achieved by surface finishes or treatments, the major item is covered with an insulating blanket composed of multiple-sheet aluminized mylar. This type of thermal control is called "passive" control.

The major items whose survival or operating temperature requirements cannot be achieved by surface finish-

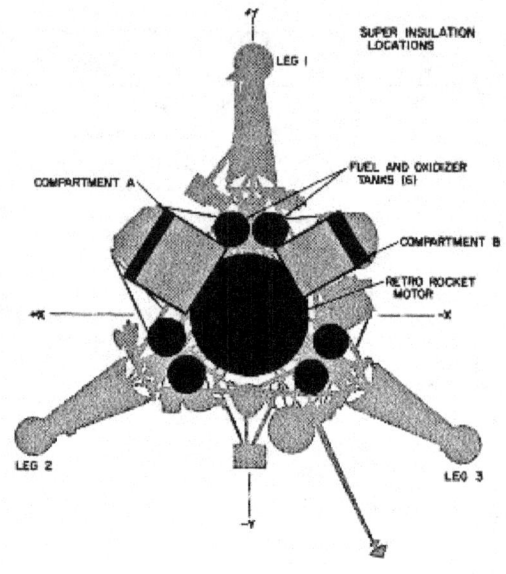

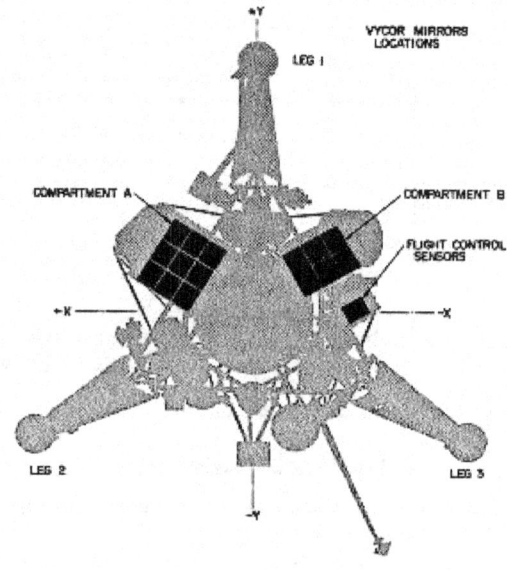

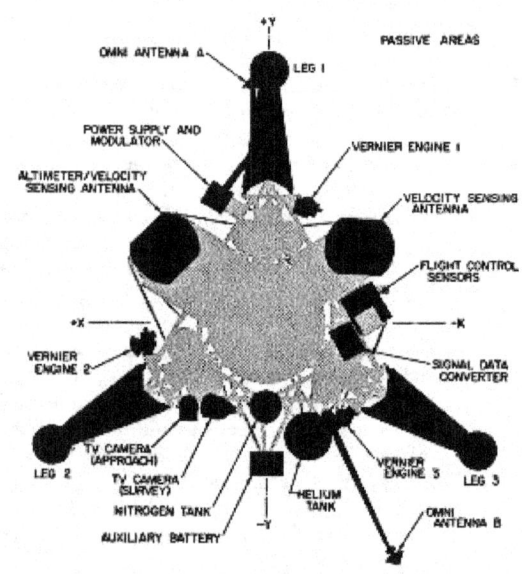

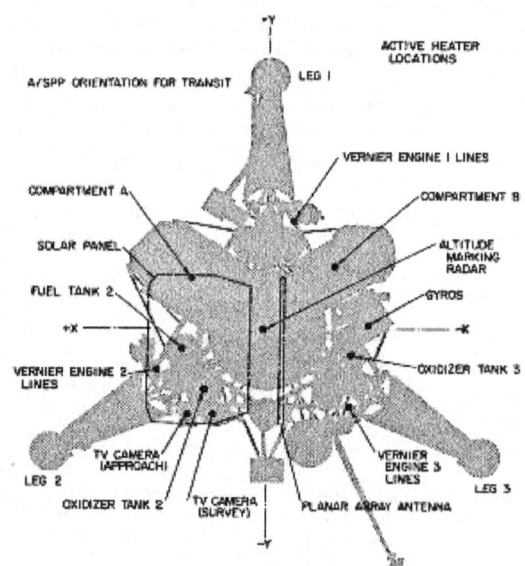

ing or insulation alone use heaters that are located within the unit. These heaters can be operated by external command, thermostatic actuation, or both. The thermal control design of those units using auxiliary heaters also includes the use of surface finishing and insulating blankets to optimize heater effectiveness and to minimize the electrical energy required. Heaters are considered "active control."

Items of electronic equipment whose temperature requirements cannot be met by the above techniques are located in thermally controlled compartments (A and B). Each compartment is enclosed by a shell covering the bottom and four sides and contains a structural tray on which the electronic equipment is mounted. The top of each compartment is equipped with a number of temperature-actuated switches (9 in Compartment A and 6 in Compartment B). These switches, which are attached

to the top of the tray, vary the thermal conductance between the tray and the outer radiator surfaces, thereby varying the heat-dissipation capability of the compartments. When the tray temperature increases, heat transfer across the switch increases. During the lunar night, the switch opens, decreasing the conductance between the tray and the radiators to a very low value in order to conserve heat. When dissipation of heat from the electronic equipment is not sufficient to maintain the required minimum tray temperature, a heater on the tray supplies the necessary heat. The switches are considered "semi-active."

Examples of units which are controlled by active, semi-active, or passive means are shown in Fig. IV-17.

The thermal performance of the spacecraft during transit was better than expected. In general, the units

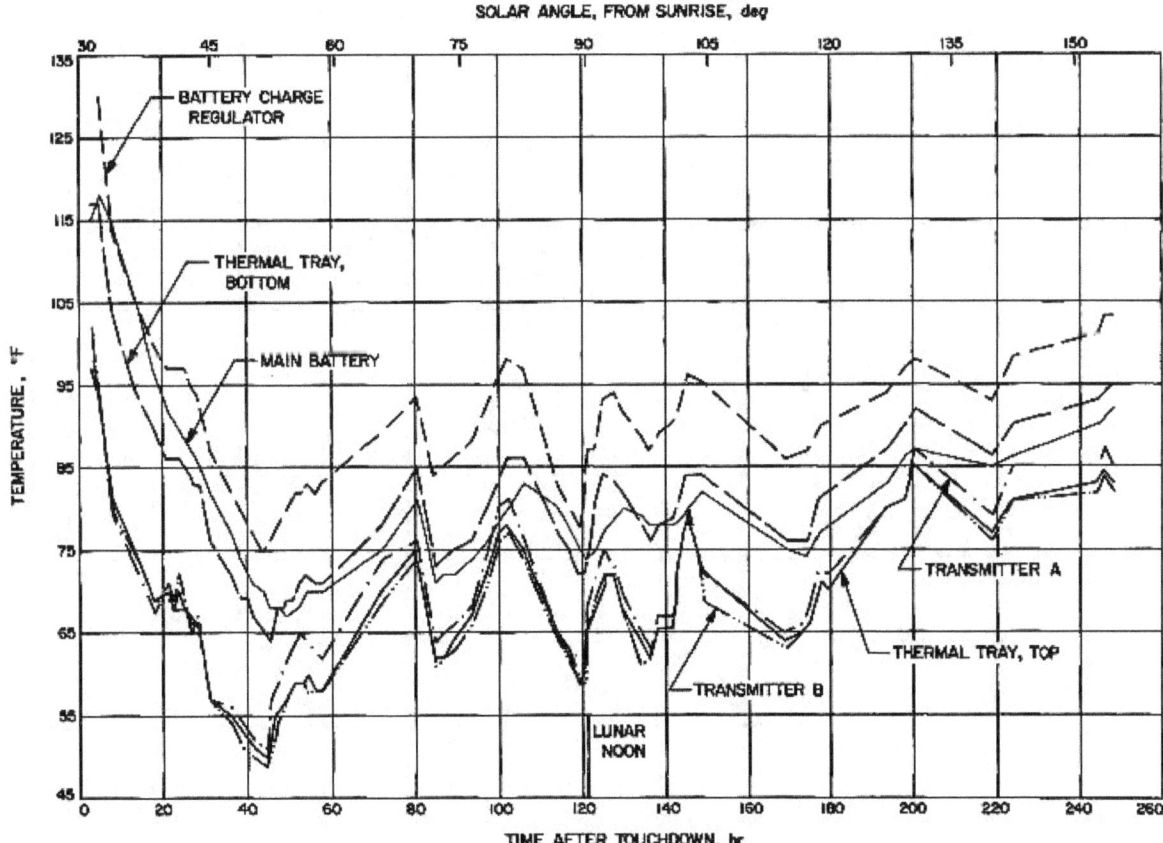

Fig. IV-18. Lunar temperatures, Compartment A

where high temperature was of concern operated at temperatures at or below the expected, while those units where low temperature was of concern operated at temperatures at or above that expected. (Appendix C presents graphs of spacecraft-component temperatures during transit.)

During transit, just prior to terminal descent, the auxiliary battery was some 25°F lower than its predicted temperature. A slight uncertainty in the absorption of the white paint used for passive thermal control is considered the reason for the lower-than-expected temperature. By judicious use of energy from the auxiliary battery, its temperature was brought to the optimum point just prior to terminal descent.

Temperature-time histories for 250 hr after touchdown for the electronics and thermal trays are shown in Figs.

IV-18 and IV-19. Compartment thermal performance was greatly enhanced by planar array/solar panel shadowing for a good part of the lunar day, as shown in Fig. IV-4. Temperatures on the transmitters, battery charge regulator (BCR), and main battery were well within design temperature limits. TV thermal performance was nominal. Temperatures on the fuel and oxidizer tanks were well below the danger point. The temperature of the engines was high enough to cause seal degradation, although no damage to the spacecraft occurred. Lunar temperature histories for Vernier Engine No. 1, Fuel Tank No. 1, Solar Panel, Transmitter A, and TV survey camera electronics are shown graphically in Appendix C.

After lunar sunset, the primary engineering experiment conducted on the spacecraft was aimed at determining the equilibrium heat loss from the compartments during the lunar night. This information was required in order to

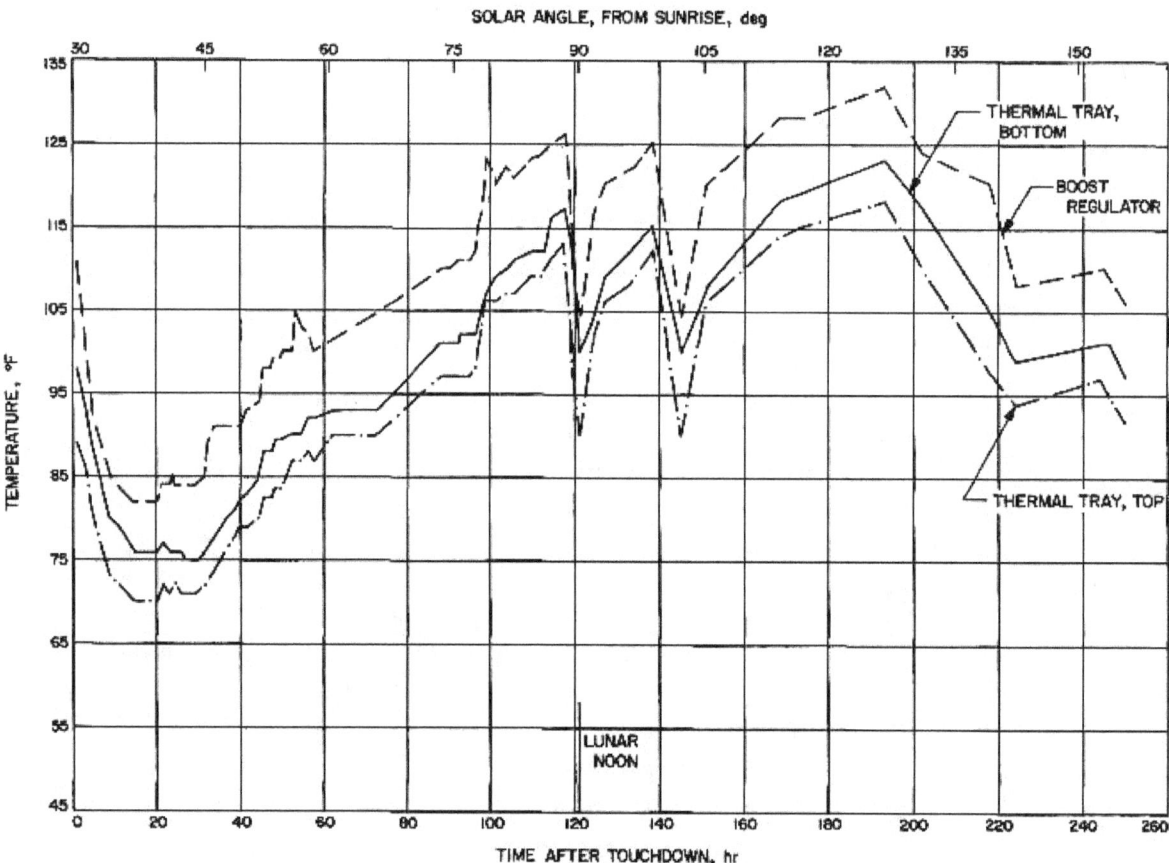

Fig. IV-19. Lunar temperatures, Compartment B

assess the capability of the spacecraft to survive the lunar night in an operational condition. Prelaunch analysis of data generated with the thermal control model (TCM) indicated a power loss larger than could be replaced with a fully charged battery.

At the time the spacecraft entered the lunar night, the power dissipation in the compartments was reduced to an absolute minimum in order to cool them as rapidly as possible. The intent was to cool them to a temperature at which the thermal switches would be opened, and then to hold at a steady temperature by periodic application of power for a time long enough to obtain a good value for steady-state power loss. This effort was hampered by the fact that several of the thermal switches did not open. Considerable effort was aimed at trying to open the stuck switches. At 2300 GMT on June 15, the last switch on Compartment A opened, while at least two remained closed on Compartment B.

The remaining time to shutdown was used to obtain data on the heat loss from Compartment A. No accurate assessment has been made because of uncertainties in the actual electrical power dissipated in the compartment. Preliminary analysis indicated that the power loss was somewhat lower than predicted, which should increase the probability of lunar-night survival.

D. Electrical Power

The electrical power subsystem is designed to generate, store, convert, and distribute electrical energy. A block diagram of the subsystem is shown in Fig. IV-20. The subsystem derives its energy from the solar panel and the spacecraft battery system. The solar panel converts solar radiation energy into electrical energy. Solar panel power capability is affected by temperature and the incidence of solar radiation and varies from 90 to 55 w.

The spacecraft battery system consists of a main battery and an auxiliary battery. The main battery is a secondary or rechargeable battery; the auxiliary battery is a primary or nonrechargeable battery.

The batteries provide about 4090 w during transit, the balance of the energy being supplied by the solar panel. The maximum storage capacity of the main battery is 180 amp-hr; that of the auxiliary battery is 50 amp-hr. The selection of battery operation mode is determined by the auxiliary battery control (ABC). There are three modes of battery operation: main battery mode, auxiliary battery mode, and high-current mode. In the main bat-

tery mode only the main battery is connected to the unregulated bus. This is the nominal configuration. In the auxiliary battery mode the main battery is connected to the unregulated bus through a series diode while the auxiliary battery is directly connected. In the high-current mode both the main and auxiliary batteries are connected to the unregulated bus without the series diode. The battery modes are changed by earth commands except that the ABC automatically switches to auxiliary battery mode from main battery mode in case of main battery failure. This automatic function can be disabled by earth command.

The four modes of solar panel operation are controlled by the battery charge regulator (BCR). In the *on* mode the optimum charge regulator (OCR) tracks the volt-ampere characteristic curve of the solar panel and hunts about the maximum power point. In the OCR *off* mode the solar panel output is switched off. This mode is intended to prevent overcharging of the main battery by the solar panel. In the OCR *bypass* mode the solar panel is connected directly to the unregulated bus. This mode is used in case of OCR failure. In the *trickle charge* mode the main battery charging current is controlled by its terminal voltage. Three BCR modes, excluding the *trickle charge* mode, are controlled by earth commands.

The OCR *off* and *trickle charge* modes are automatically controlled by the battery charge logic circuitry (BCL). When the main battery terminal voltage exceeds 27.5 v or its manifold pressure exceeds 60 psia, the BCR goes automatically to the *off* mode. The *trickle charge* mode is automatically enabled when the main battery terminal voltage reaches 27.3 v. The BCL can be disabled by earth command.

Current from the BCR and spacecraft batteries is distributed to the unregulated loads and the boost regulator (BR) via the unregulated bus. The voltage on the unregulated bus can vary between 17.5 and 27.5 v, with a nominal value of 22 v. The BR converts the unregulated bus voltage to 29.0 v $\pm 1\%$ and supplies the regulated loads. The preregulator supplies a regulated 30.4 v dc to the preregulated bus. The essential loads are fed by the preregulated bus through two series diodes. The diodes drop the preregulated bus voltage of 30.4 v to the essential bus voltage of 29.0 v. The preregulated bus also feeds the flight control regulator and the nonessential regulator, which in turn feeds the flight control and nonessential busses. These regulators can be turned on and off by earth commands. The nonessential regulator has a bypass mode of operation which connects the preregulated bus

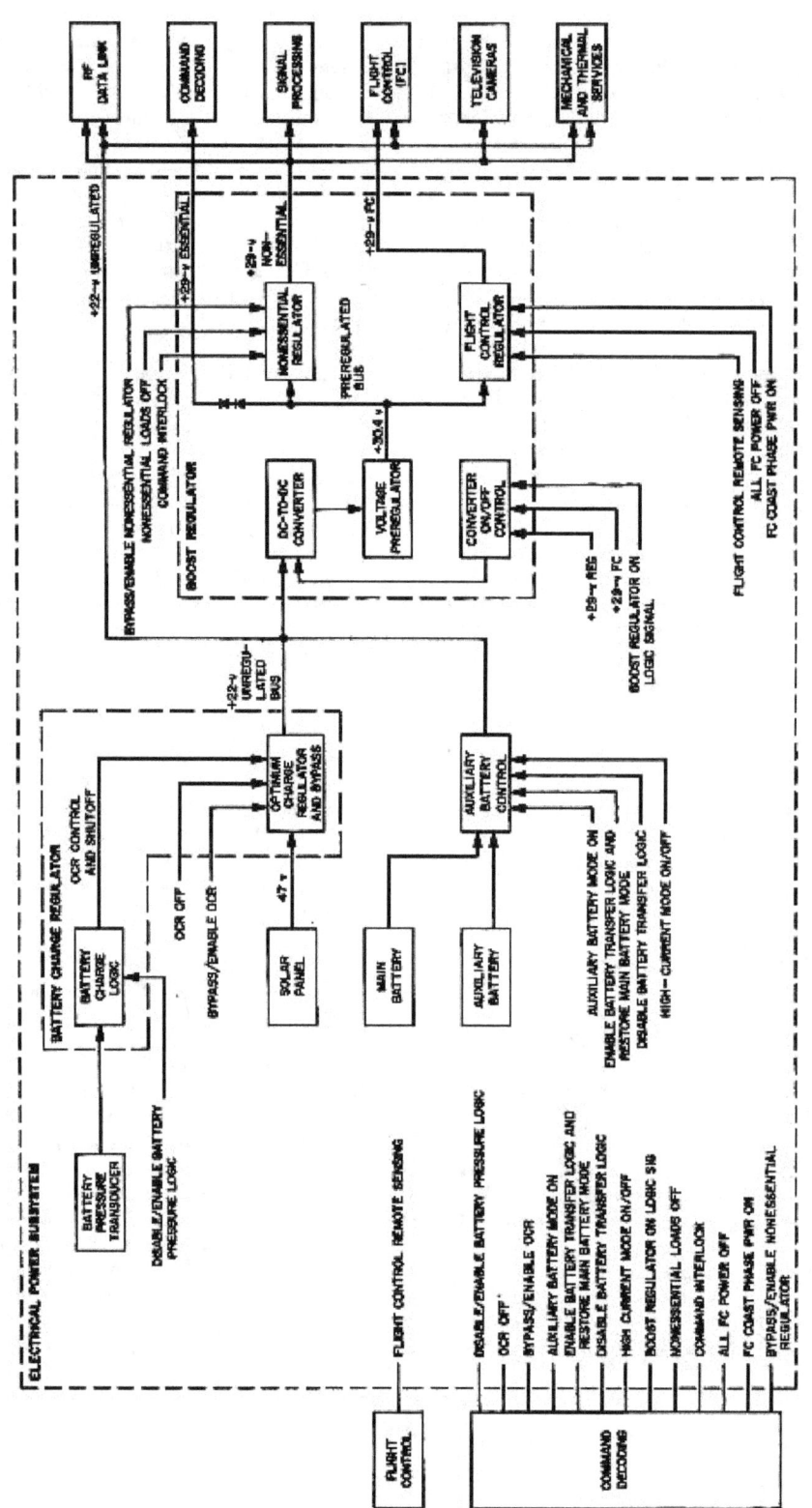

Fig. IV-20. Simplified electrical power functional block diagram

Table IV-10. Comparison of predicted vs actual values

Phase	Battery discharge current, amps		Regulated current, amps		OCR output current, amps		Unregulated current, amps	
	Predicted	Actual	Predicted	Actual	Predicted	Actual	Predicted	Actual
High power after separation and sun acquisition	5.7	5.20	4.60	4.40	3.06	3.08	0.85	0.70
First coast phase	1.63	1.74	2.34	2.39	3.14	3.15	0.86	0.81
Second coast phase	1.6	1.85	2.31	2.38	3.14	3.15	0.86	0.80

directly to the nonessential bus. This mode is used if the nonessential regulator fails.

The power subsystem operated normally throughout the mission. Table IV-10 verifies that telemetered parameters were in close agreement with the predicted values.

During Coast Phase I the average regulated load was 2.39 amp (Fig. IV-21), with an average BR efficiency of 79%, and the average unregulated current was 0.81 amp (Fig. IV-22). Comparable time period predictions indicate that the regulated load should be 2.34 amp and the unregulated current 0.80 to 0.83 amp. During low-power transmitter interrogation, the regulated output current was approximately 50-100 ma higher than predicted. During high-power transmitter interrogation, this current read 200 to 300 ma low.

For the above mission period, the average (OCR) output current was 3.15 amp (Fig. IV-23), which agrees closely with test data. The average solar panel output current was 1.8 amp (Fig. IV-24) at an average voltage of 48.5 (Fig. IV-25). The overall OCR efficiency was about 79%. The OCR solar panel combination supplied an average of 66% of the total system electrical loads, with the battery providing the remaining 34% of the load.

Battery pressure stabilized during this period to 18 psi (Fig. IV-26) at a steady-state battery temperature of 96°F, both measurements falling well within the normal safe operating limits. Main battery terminal voltage and discharge current are shown in Figs. IV-27 and IV-28. The auxiliary battery voltage history is shown in Fig. IV-29. Figures IV-30, IV-31, and IV-32 show the BR pre-regulator, 29-v nonessential, and unregulated bus voltages vs mission time. The actual and predicted battery energy

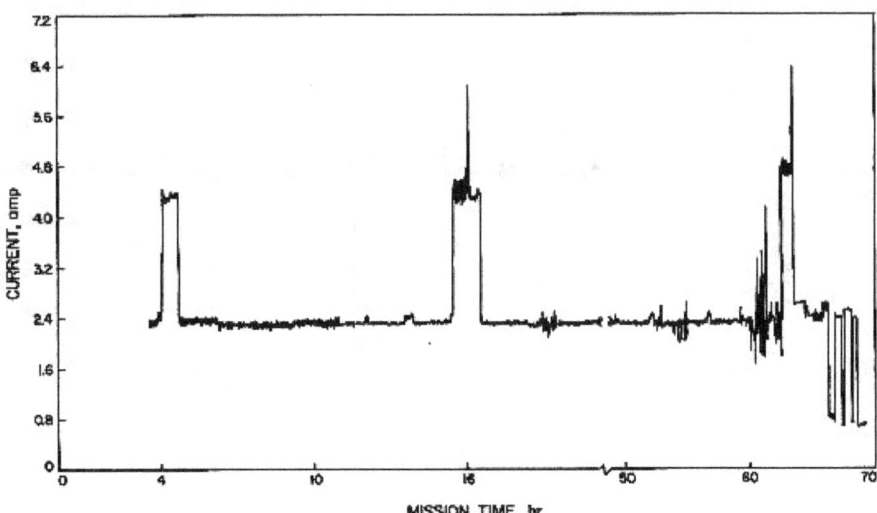

Fig. IV-21. Regulated output current

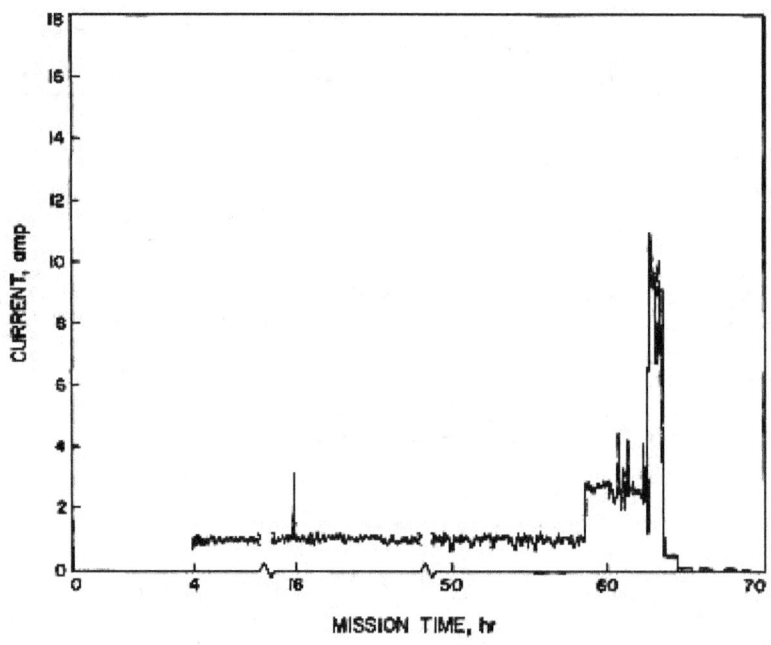

Fig. IV-22. Unregulated output current

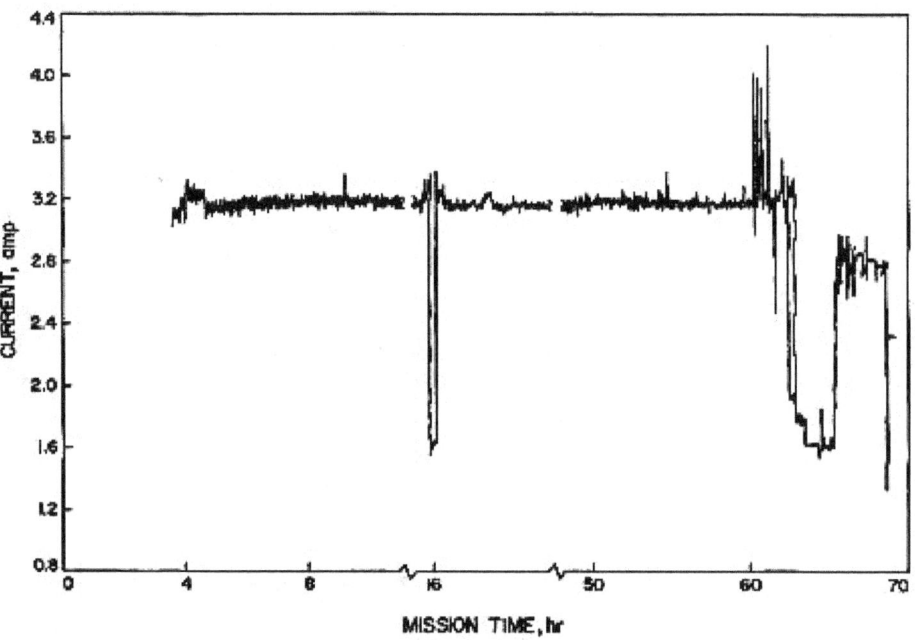

Fig. IV-23. OCR output current

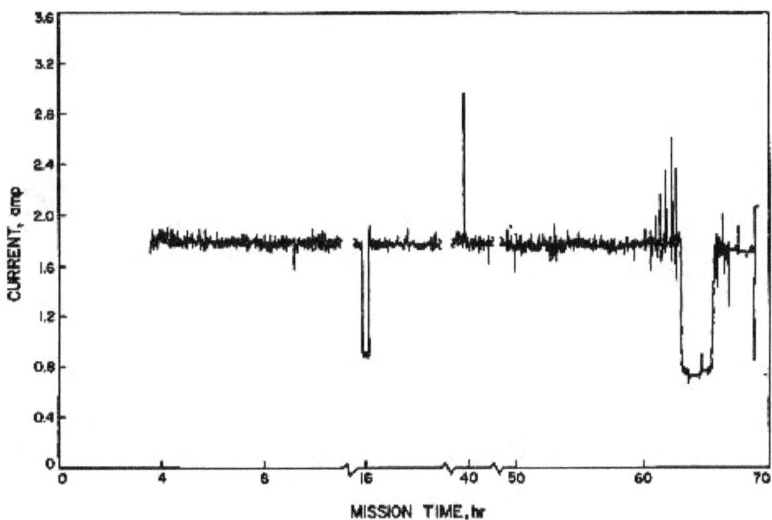

Fig. IV-24. Solar cell array current

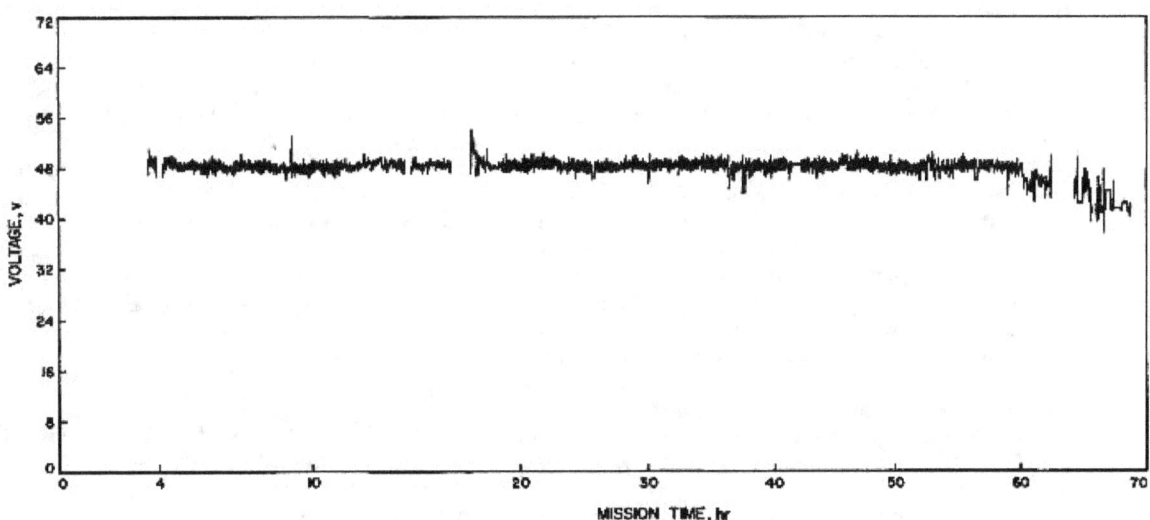

Fig. IV-25. Solar cell array voltage

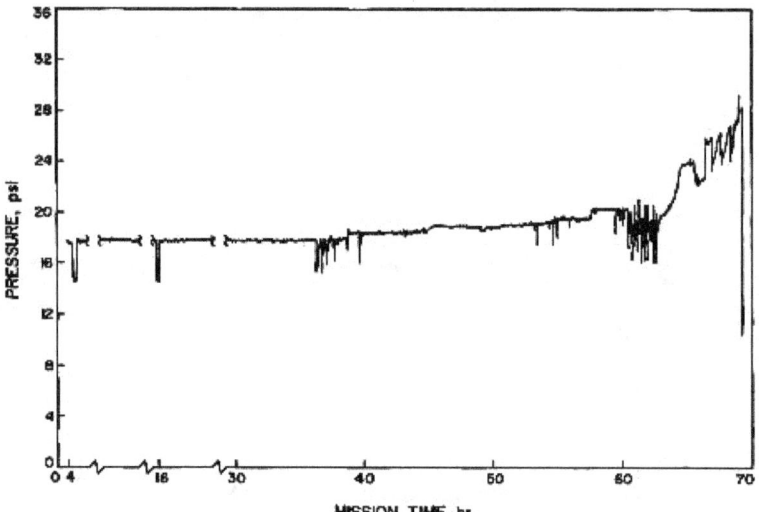

Fig. IV-26. Main battery manifold pressure

profiles are shown in Fig. IV-33. The prelaunch pre-dictions were revised from 70.4 to 72.7 amp-hr during the mission for the battery energy (main and auxiliary) remaining at touchdown. This increase was due to the elimination of several high-power interrogations. The flight analysis of the energy available at touchdown (TD) was 74 ±15 amp-hr. Figure IV-33 also shows the power consumption history for the transit phase (predicted and actual).

During the last part of the transit, the auxiliary battery temperature was lower than expected and this condition could have prevented the auxiliary battery from maintain-ing an adequate terminal voltage during terminal descent in the event of main battery failure. The auxiliary battery temperature was raised by operating in auxiliary battery mode until TD-5 hr. This also allowed the main battery to retain a greater amount of energy at touchdown, though at the expense of the energy margin planned for the auxiliary battery. The auxiliary battery temperature rose from 35°F at 45 hr after launch (L + 45 hr) to 64°F at L + 61 hr and to 77.8°F at L + 64.8 hr (post-touchdown). The auxiliary battery remained below its temperature limit of 125°F after touchdown, and the high-current mode was maintained until L + 66.3 hr. At this time, the transmitter high-power mode was turned off and main battery charging began, requiring the removal of the auxiliary battery from the line. At TD + 8 days, the auxiliary battery continued to maintain a terminal volt-age (open circuit) of approximately 22.4 v.

In order to reduce thermal dissipation in Compart-ment A and to reduce the main battery temperature during the post-touchdown phase, the power subsystem was operated in OCR bypass mode. The main battery was charged at approximately 1 amp in this mode, with an average solar panel voltage of 29 v. On L + 218 hr the charge current was reduced to prevent exceeding 27.3 v terminal voltage. During the initial charge period the battery pressure rose slowly to about 37 psi, where it stabilized. With the battery recharged to almost a full-charged condition, the main battery energy was approx-imately 162 amp-hr at the time of the day/night terminator. Later, when spacecraft operations were dis-continued for the duration of the lunar night, the remaining battery energy was estimated to be 107.5 amp-hr.

E. Flight Control

The flight control subsystem (FCS) controls the space-craft velocity and attitude during transit from the time of spacecraft separation from the *Centaur* vehicle to spacecraft touchdown on the lunar surface. The basic functions performed by the FCS include:

1. Attitude stabilization and orientation during transit.

2. Midcourse trajectory correction based on radio command data.

3. Retro pyrotechnic and vernier descent control for soft-landing the spacecraft.

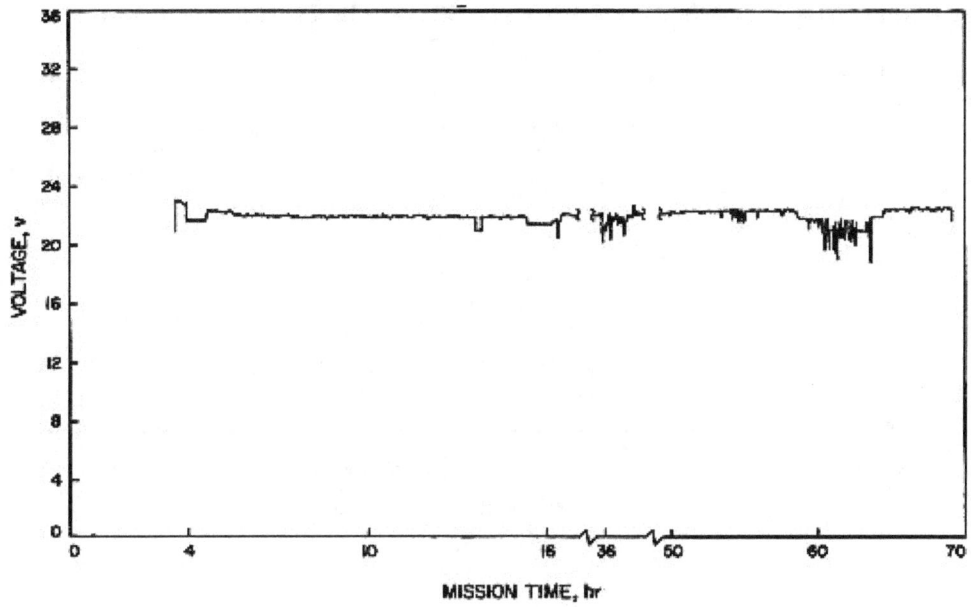

Fig. IV-27. Main battery voltage

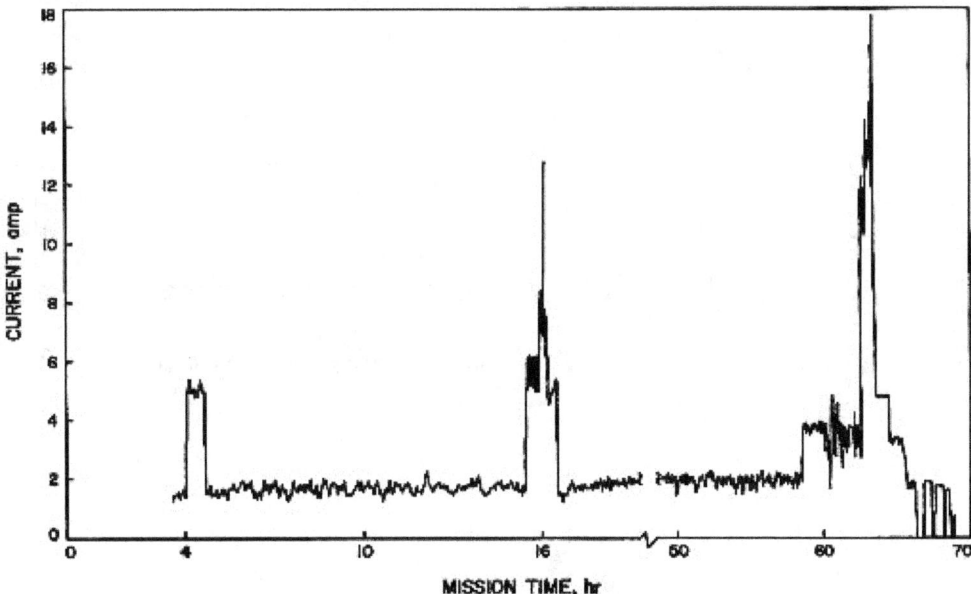

Fig. IV-28. Main battery discharge current

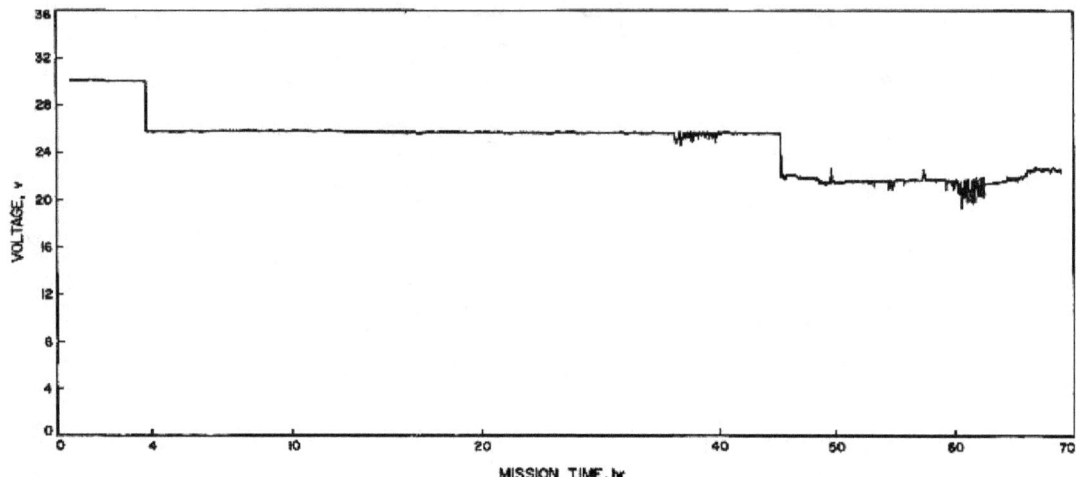

Fig. IV-29. Auxiliary battery voltage

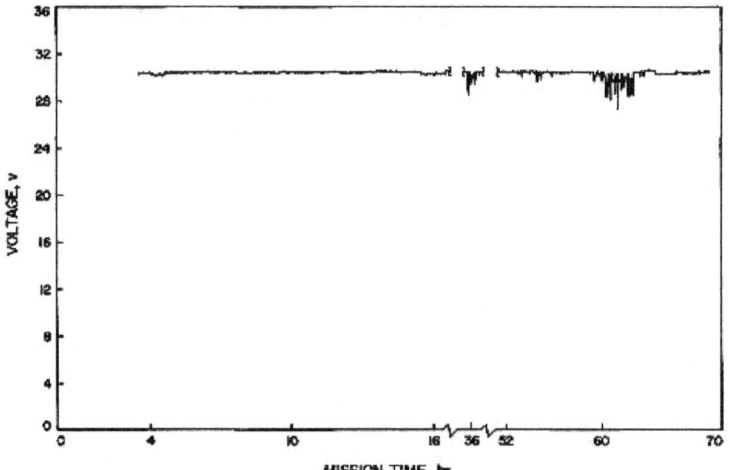

Fig. IV-30. Boost regulator preregulator voltage

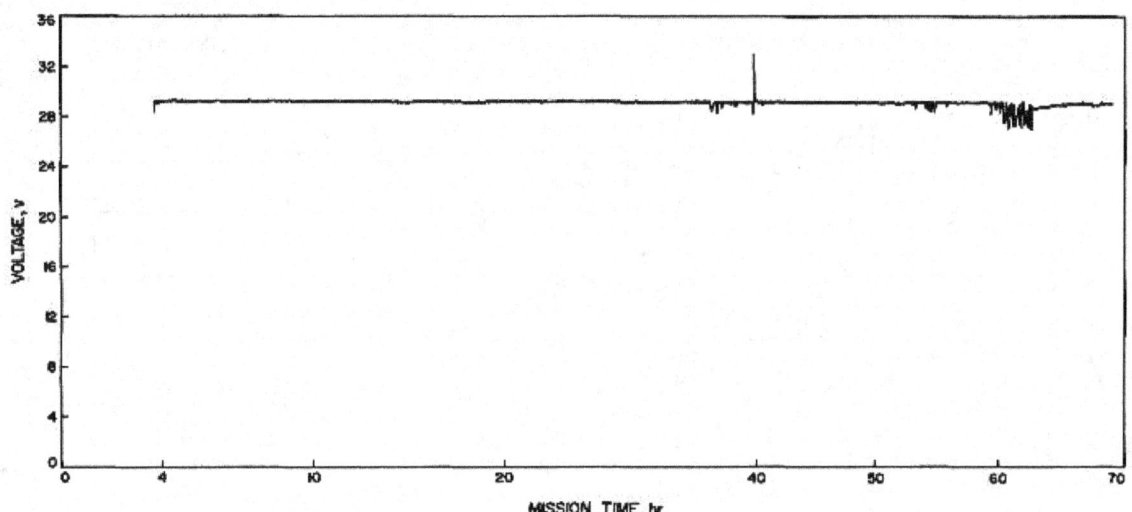

Fig. IV-31. 29-v unregulated bus voltage

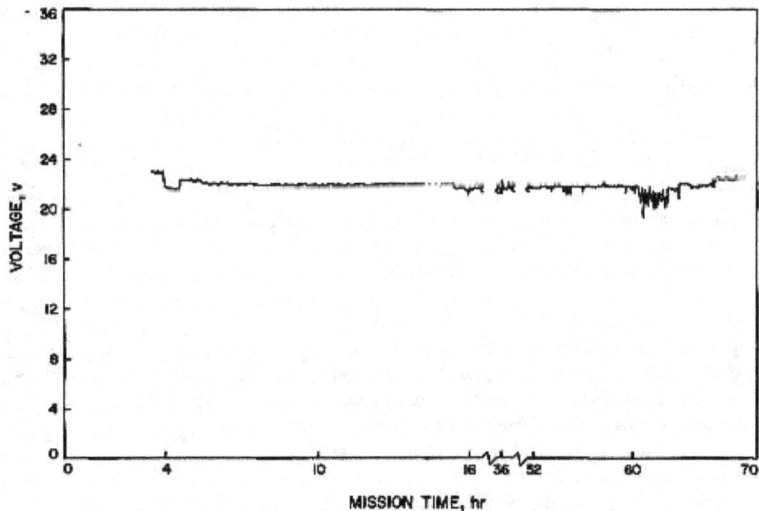

Fig. IV-32. 22-v unregulated bus voltage

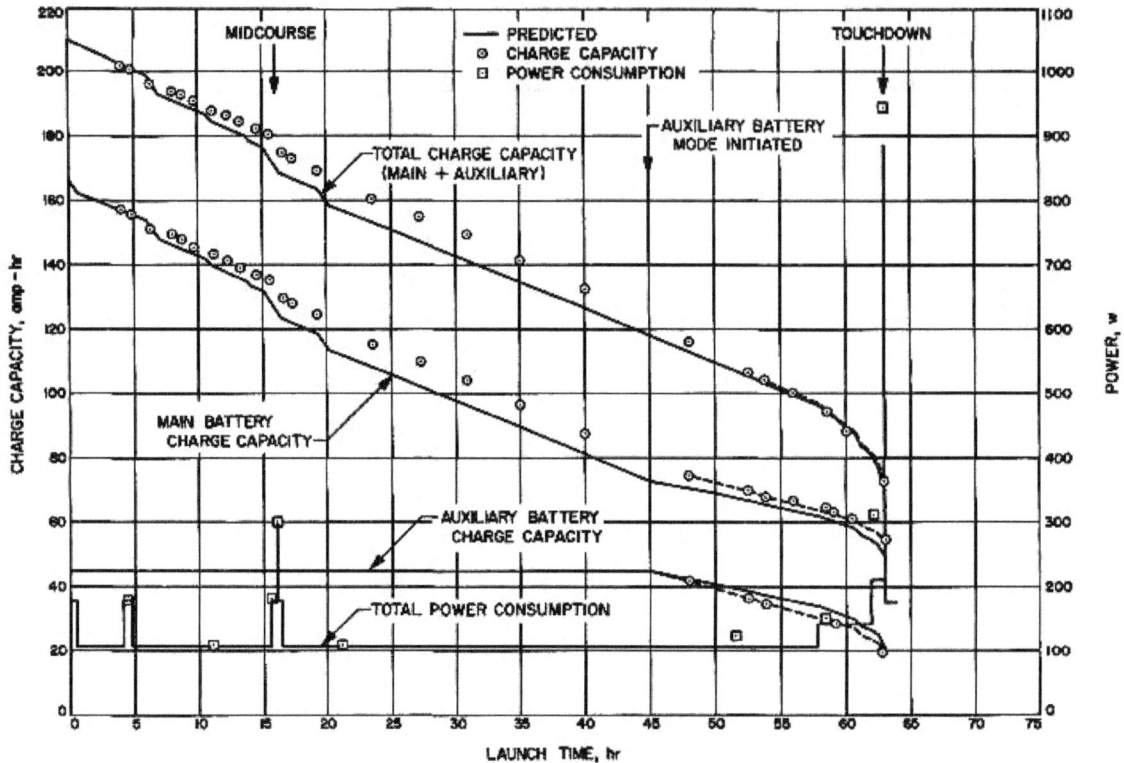

Fig. IV-33. Battery energy capacity and total power consumption profile

The FCS consists of reference sensing elements, flight control and mode control electronics, and vehicle control elements functionally arranged as shown in Fig. IV-34.

The principal in-flight references used by the spacecraft are inertial, celestial, and lunar; each is sensed respectively by inertia, optical, and radar sensors. The control electronics process the reference sensor outputs, earth-based commands, and the on-board flight control programmer and decoder outputs to generate the necessary control signals for use by the vehicle control elements. The vehicle control elements consist of the attitude-control cold gas jet activation valves and gas supply system, the vernier engine throttleable thrust valve and controllable gimbal actuator, and the main retro igniter and retro case separation pyrotechnics.

Vehicle response in attitude, acceleration, and velocity is controlled as needed by various "control loops"

throughout the coast and thrust phases of flight, as shown in Table IV-11. Stabilization of the spacecraft tipoff rates after Centaur separation is achieved through the use of rate feedback gyro control (rate mode). After rate capture, an inertial mode is achieved by switching to position feedback gyro control. Because of the long duration of the transit phase and the small unavoidable drift error of the gyros, a celestial reference is used to continuously update the inertially controlled attitude of the spacecraft.

The celestial references (Fig. IV-5), the sun and the star Canopus, are acquired and maintained automatically after the spacecraft separates from the Centaur stage and after automatic deployment of the solar panel. The sun is first acquired by the automatic sun sensor during an automatically commanded spacecraft roll maneuver. The 10-deg wide × 196-deg fan-shaped field of view of the automatic sun sensor includes the Z axis and is centered

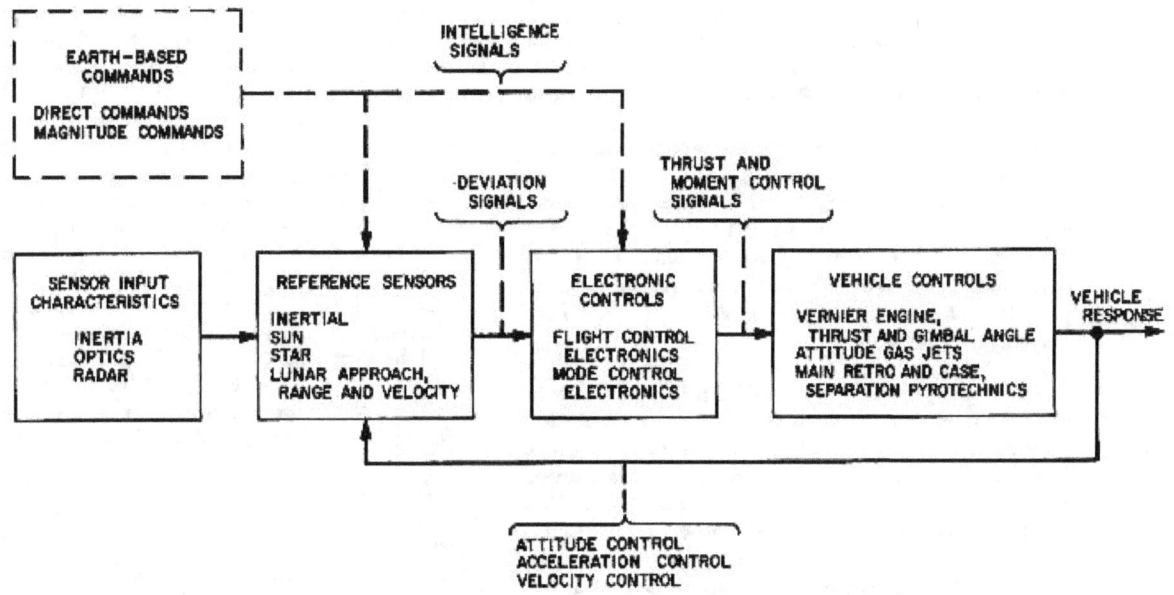

Fig. IV-34. Simplified flight control functional block diagram

Table IV-11. Flight control modes

Control loop	Flight phase	Modes	Remarks
Attitude control loop			
Pitch and yaw	Coast	Rate Inertial Celestial	Gas jet matrix signals
	Thrust	Inertial Lunar radar	Vernier engine matrix signals
Roll	Coast	Rate Inertial Celestial	Leg No. 1 gas jet signals
	Thrust	Inertial	Vernier engine No. 1 gimbal command
Acceleration control loop			
Thrust axis	Thrust (midcourse)	Inertial (with accelerometer)	Nominal 3.22 ft/sec^2
	Thrust (terminal descent)	Inertial (with accelerometer)	Minimum 4.77 ft/sec^2
			Maximum 12.56 ft/sec^2
Velocity control loop			
Thrust axis	Thrust	Lunar radar	Command segment signals to 43 ft altitude
			Constant 5-fps velocity signals to 14-ft altitude
Lateral axis	Thrust	Lunar radar	Lateral/angular conversion signals

about the minus X axis. The roll command is terminated after initial sun acquisition, and a yaw command is initiated which allows the narrow-view primary sun sensor to acquire and lock on the sun. Automatic Canopus acquisition and lock-on are then achieved after initiation of a roll command from earth. This occurs because the Canopus sensor angle is preset with respect to the primary sun sensor prior to launch for each mission. Star mapping for Canopus verification is achieved by commanding the spacecraft to roll while the spacecraft maintains sun lock. A secondary sun sensor, mounted on the solar panel, provides a backup for manual acquisition of the sun if the automatic sequence fails.

The transit phase is performed with the spacecraft in the celestial-referenced mode except for the initial rate-stabilization, midcourse, and terminal descent maneuvers. The midcourse and main retro orientation maneuvers are achieved in the inertial mode. Acceleration control is used for controlling the magnitude of the midcourse velocity increment. During the interval from retro case separation to initiation of radar velocity control, acceleration control is used to control the descent along the spacecraft thrust axis, and velocity control is used for pitch and yaw control to align the spacecraft thrust axis with the velocity vector.

The lunar reference is first established by a signal from an altitude marking radar (AMR) subsystem when the spacecraft is nominally 60 miles above the lunar surface. Dual-channel video gating (early and late gating signals) is used. The early and late gates are adjacent at 60 miles so that as the spacecraft approaches the lunar surface, the video return becomes equally coincident with these two gates, making it possible to mark accurately on the center of the RF return. When the center of the main lobe is at a slant range of 60 miles, the altitude mark signal is generated. The AMR signal initiates automatic terminal descent spacecraft operations.

Subsequent lunar reference in the form of continuous range and velocity information is provided by a 4-beam RADVS subsystem.

1. Radar Altimeter and Doppler Velocity Sensor (RADVS)

The RADVS (Fig. IV-35) functions in the flight control subsystem to provide three-axis velocity, range, and altitude mark signals for flight control during the retro and vernier phases. The RADVS consists of a doppler velocity sensor (DVS), which computes velocity along the spacecraft X, Y, and Z axes, and a radar altimeter (RA), which computes slant range from 40,000 to 14 ft and generates 1000-ft and 14-ft mark signals. The RADVS comprises four assemblies: (1) klystron power supply/modulator (KPSM), which contains the RA and DVS klystrons, klystron power supplies, and altimeter modulator, (2) altimeter/velocity sensor antenna, which contains beams 1 and 4 transmitting and receiving antennas and preamplifiers, (3) velocity sensing antenna, which contains beams 2 and 3 transmitting antennas and preamplifiers, (4) RADVS signal data converter, which consists of the electronics to convert doppler shift signals into dc analog signals. The RADVS is turned on at about 50 miles above the lunar surface and is turned off at about 13 ft. Fig. IV-36 shows the RADVS 4-beam configuration.

a. Doppler velocity sensor. The doppler velocity sensor (DVS) operates on the principle that a reflected signal has a doppler frequency shift proportional to the approaching velocity. The reflected signal frequency is higher than the transmitted frequency for the closing condition. Three beams directed toward the lunar surface enable velocities in an orthogonal coordinate system to be determined.

The KPSM provides an unmodulated DVS klystron output at a frequency of 13.3 kmc. This output is fed equally to the DVS1, DVS2, and DVS3 antennas. The RADVS velocity sensor antenna unit and the altimeter velocity sensor antenna unit provide both transmitting and receiving antennas for all three beams. The reflected signals are mixed with a small portion of the transmitted frequency at two points ¾ wavelength apart for phase determination, detected, and amplified by variable-gain amplifiers providing 40, 65, or 90 db of amplification, depending on received signal strength. The preamp output signals consist of two doppler frequencies, shifted by ¾ transmitted wavelength, and preamp gain-state signals for each beam. The signals are routed to the trackers in the RADVS signal data converter.

The D1 through D3 trackers in the signal data converter are similar in their operation. Each provides an output which is 600 kc plus the doppler frequency for approaching doppler shifts. If no doppler signal is present, the tracker will operate in search mode, scanning frequencies between 82 kc and 800 cps before retro burnout, or between 22 kc and 800 cps after retro burnout. When a doppler shift is obtained, the tracker will

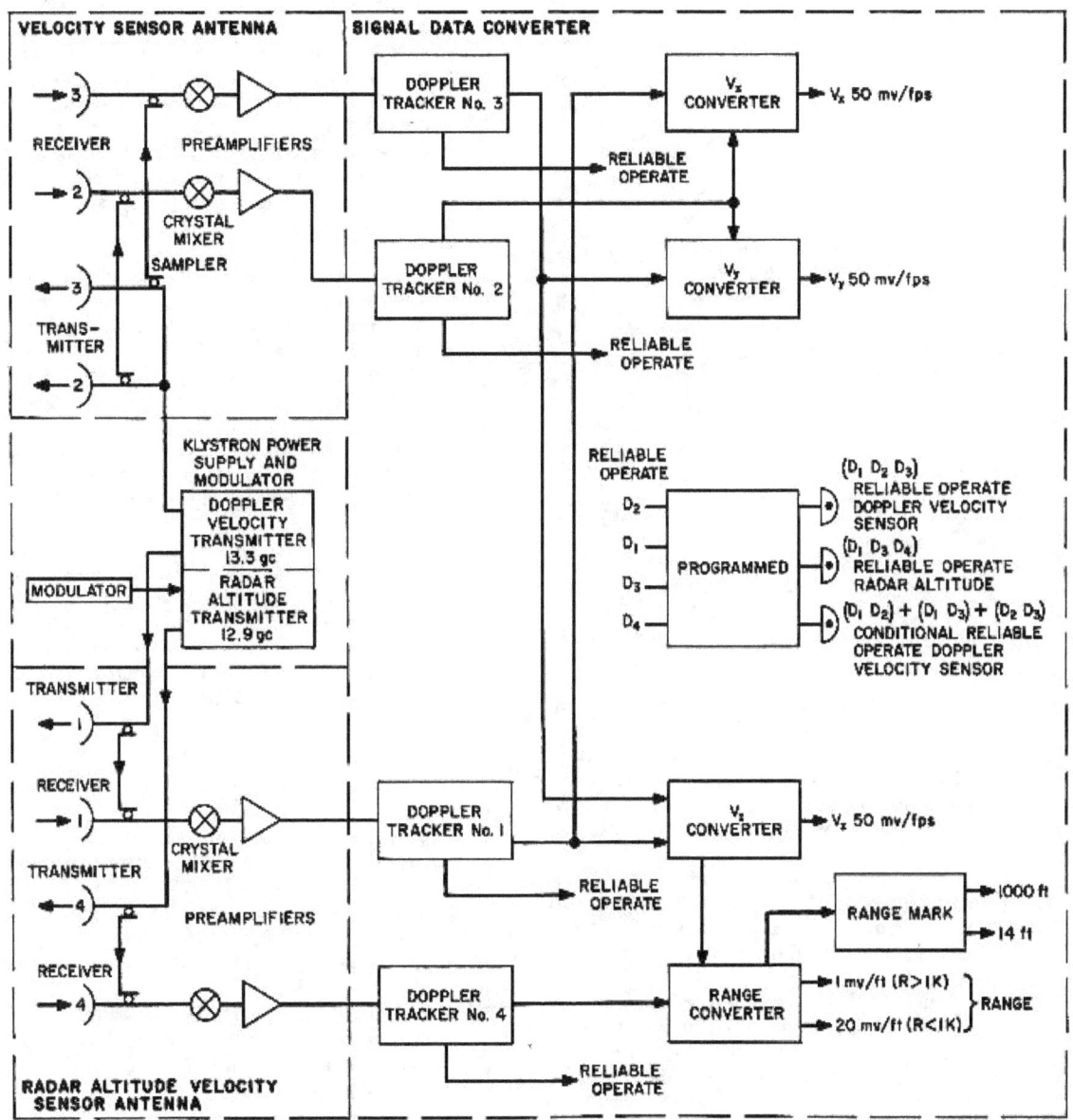

Fig. IV-35. Simplified RADVS functional block diagram

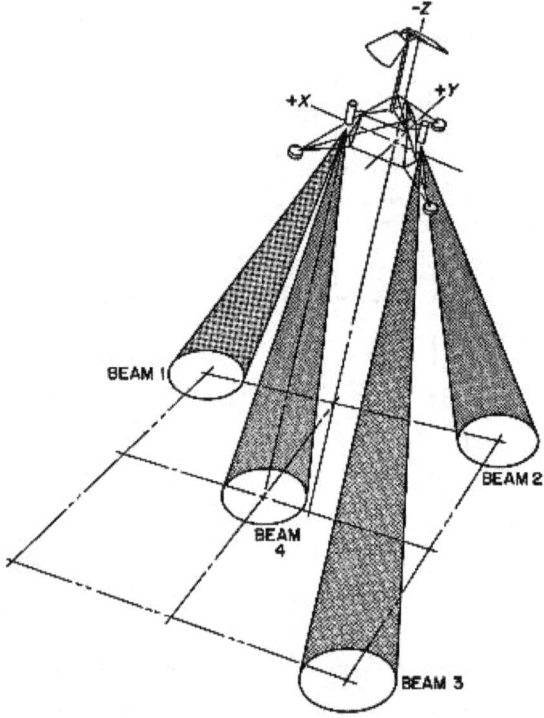

Fig. IV-36. RADVS beam orientation

The RF signal is radiated, and the reflected signal is received by the altimeter/velocity sensor antenna. The received signal is mixed with two samples of transmitted energy ¾ wavelength apart, detected, and amplified by 40, 65, or 90 db in the altimeter preamp, depending on signal strength. The signals produced are difference frequencies resulting from the time lag between transmitted and received signals of a known shift rate, coupled with an additional doppler frequency shift because of the spacecraft velocity.

The altimeter tracker in the signal data converter accepts doppler shift signals and gain-state signals from the altimeter/velocity sensor antenna and converts these into a signal which is 600 kc plus the range frequency plus the doppler frequency. This signal is routed to the altimeter converter for range dc analog signal generation.

The range mark, reliability, and reference circuits produce the *1000-ft mark* signal and the *14-ft mark* signal from the *range* signal generated by the altimeter converter.

The *range mark* and *reliable* signals are routed to flight control electronics. The signals are used to rescale the *range* signal, for vernier engine shutoff and to indicate whether or not the *range* signal is reliable. The *reliable* operate signal is also routed to signal processing for transmission to DSIF.

operate as described above and initiate a lock-on signal. The tracker also determines amplitude of the reflected signal and routes this information to the signal processing electronics for telemetry.

The velocity converter combines tracker output signals D_1 through D_3 to obtain dc analog signals corresponding to the spacecraft X, Y, and Z velocities; $D_1 + D_3$ is also sent to the altimeter converter to compute range.

Range mark, reliability, and reference circuits produce a *reliable operate* signal if D_1 through D_3 lock-on signals are present, or if any of these signals are present 3 sec after retro burnout. The *reliable operate* DVS signal is routed to the flight control electronics and to the signal processing electronics telemetry.

b. Radar altimeter. Slant range is determined by measuring the reflection time delay between the transmitted and received signals. The transmitted signal is frequency-modulated at a changing rate so that return signals can be identified.

2. Terminal Descent Design

The terminal maneuver is initiated by pointing the vehicle thrust axis in a direction which is precalculated on earth to be nearly aligned with the predicted velocity vector at main retro ignition. Then, when the distance to lunar surface reaches a preset value (about 60 miles), a pulse-type radar altimeter (AMR) generates a marking signal. After a suitable time delay, precomputed on earth, a solid-propellant main retro engine is ignited.

During main engine burning, the vehicle attitude is maintained constant by differential throttling of the three controllable liquid-propellant engines used previously for the midcourse maneuver. The solid motor burns at an essentially constant thrust for about 40 sec after which the thrust starts to decay. This tailoff is detected by an acceleration switch which initiates a time delay of about 8 sec, after which the empty main retro engine case is jettisoned. The remainder of the descent is made with the vernier engines, commanded according to on-board measurements of altitude and velocity.

The vernier phase generally begins at altitudes between 10,000 and 50,000 ft and velocities in the range of 100 to 700 ft/sec. Because of statistical variations in various parameters, a substantial range of burnout conditions is possible. The vehicle attitude and thrust level during the vernier engine phase are controlled according to on-board (RADVS) measurements of velocity and range to the lunar surface.

Shortly after separation of the main retro case, the thrust is reduced to follow a commanded thrust acceleration of 0.9 lunar g, sensed by an axially oriented accelerometer. The spacecraft attitude is held in its original position until the velocity radar locks on to the lunar surface. The thrust axis is then aligned with the velocity vector and maintained there throughout the remainder of the descent. The vehicle descends with the thrust acceleration at 0.9 g until the radars sense that the "descent contour" has been reached (Fig. IV-37). This contour, in the range-velocity plane, is a straight-line approximation to a parabola which corresponds in the vertical case to descent at a constant acceleration based on the maximum thrust capability of the vernier engines. The thrust is then commanded such that the vehicle follows the descent contour until shortly before touchdown, when the terminal sequence is initiated. This final phase consists nominally of a constant velocity descent from 40 to 13 ft at 5 ft/sec, followed by a free fall from 13 ft, resulting in touchdown at approximately 13 ft/sec.

The horizontal component of the landing velocity is nominally zero. Dispersions arise, however, primarily because of the following two factors:

1. Measurement error in the doppler system resulting in a velocity error normal to the thrust axis.

2. Nonvertical attitude due to:
   a) Termination of the gravity turn at a finite velocity.
   b) Attitude transients resulting from radar and other control system noise sources.

Since the attitude at the beginning of the constant velocity descent is inertially held throughout the remainder of powered descent until cutoff, these errors give rise to a significant lateral velocity at touchdown. The capability to withstand this velocity without toppling is an essential requirement in the vehicle landing gear design.

Constraints on the allowable burnout region are imposed by the doppler radar and the altimeter. Linear operation of the doppler velocity sensor is expected for

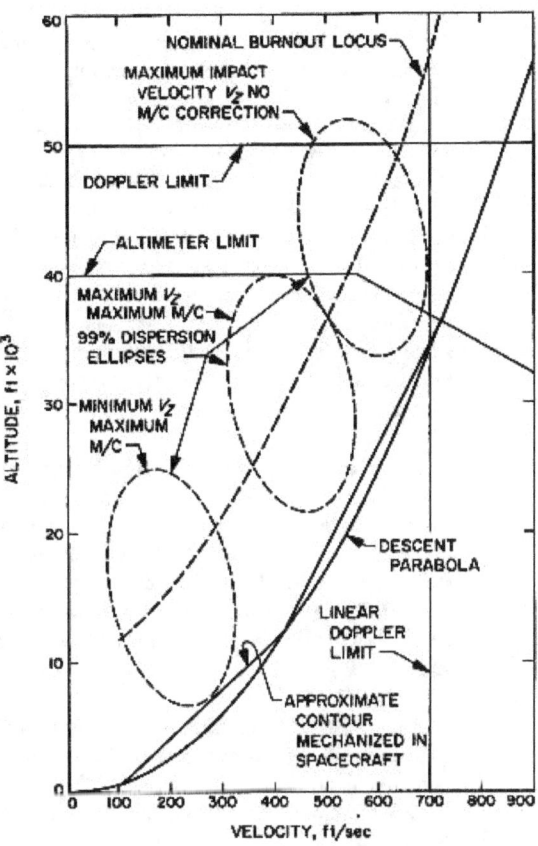

Fig. IV-37. Altitude-velocity diagram

slant ranges below 50,000 ft, and for velocities below 700 ft/sec. The altimeter limit is between 30,000 and 40,000 ft, depending on velocity. These constraints are illustrated in the range-velocity plane in Fig. IV-37.

Furthermore, satisfactory operation of the doppler radar can only be guaranteed when the flight path at burnout is within 45 deg of the vertical and when the vehicle roll axis is within 65 deg of the velocity vector. These operating constraints affect the minimum allowable burnout velocity.

The allowable burnout velocity range is further restricted by the maximum thrust capability of the vernier engine system. To accurately control the final descent, the minimum thrust must be less than the least possible landed weight of the vehicle. The result is a minimum thrust of 90 lb and a maximum of 312 lb. Descent at the maximum thrust to touchdown defines a curve in the range-velocity plane below which main retro burnout cannot be allowed to occur. Actually, since the vernier engines are also used for attitude control, it is necessary to allow some margin from the maximum thrust level. Furthermore, since it is more convenient to sense acceleration than thrust, the terminal maneuver is performed at a nominally constant acceleration rather than at constant thrust. This acceleration is determined from the maximum thrust capability of the engines and the maximum possible vehicle weight at main engine burnout, with suitable allowance for attitude control and control system mechanization errors.

This maximum commanded thrust acceleration defines a parabola in the altitude velocity plane. For vertical descents at least, this curve defines the minimum altitude at which main retro burnout is permitted to occur and still achieve a soft landing. This parabola is indicated in Fig. IV-37. For ease of spacecraft mechanization, the parabola is approximated by straight line segments, also shown in Fig. IV-37.

The main retro ignition altitude is established such that burnout will occur sufficiently above the descent contour to allow time to align the thrust axis with the velocity vector before the trajectory intersects the contour. Thus, a "nominal burnout locus" is established which allows for altitude dispersions plus an alignment time which depends on the maximum angle between the flight path and roll axis at burnout. This locus is included in Fig. IV-37.

The allowable burnout region having been defined, the size of the main retro engine is determined such that burnout will occur within that region. It is necessary to provide for dispersions in engine characteristics, as well as for vehicle weight variations due to fuel expended in performing the midcourse maneuver, and for the required range of approach velocities.

The maximum vernier fuel expenditure occurs when the fuel consumed at midcourse is also maximum. The design chosen provides enough fuel so that, given a maximum midcourse correction, the probability of not running out is at least 0.99. One way of doing this is to examine the fuel required for points on the 99% dispersion ellipse and to provide enough fuel for the worst case.

The principal sources of velocity dispersion are the imperfect alignment of the vehicle prior to retro ignition and the variability of the total impulse. These variations, in the case of a nominally vertical descent, cause dispersions of the type shown in Fig. IV-37, where the ellipse defines a region within which burnout will occur with probability 0.99.

The basic closed-loop guidance method used during the vernier phase consists of (1) continual alignment of the thrusting direction antiparallel to the instantaneous velocity vector and (2) thrust acceleration control in accordance with a nominal required velocity vs slant range "descent contour." The attitude control law (1), known as the "gravity turn," begins after the doppler radar has acquired the velocity vector and the initial thrust axis pointing error has been corrected. Regardless of the acceleration level, the gravity turn tends to force the flight path towards the vertical as time progresses. It has the very desirable property that, when the velocity becomes zero, the flight path and therefore the vehicle thrust attitude are concurrently vertical. The thrust control law (2) insures that the cutoff velocity of 5 ft/sec is reached at some suitably low altitude above the lunar surface.

3. Performance

a. Prelaunch and launch. The attitude control cold gas system was pressurized to 4650 psi at 78°F (4.49 lb of N_2) prior to launch. The roll, pitch, and yaw gyro temperatures are shown in Table IV-12 for the mission phase. All temperatures were nominal.

Table IV-12. Gyro temperatures

Time, GMT	Temperature, °F		
	Roll	Pitch	Yaw
14:22:00	167.2	172	176.3
14:41:01		Liftoff	
15:03:00	170.9	177	182.6
15:55:00	178.1	184	190
16:60:00	178.1	184	190
00:58:24	178.1	184	190
06:04:05	172.7	178	182
07:29:33	172.7	178	182
13:15:49	172.7	178	182.6

b. Separation and coast phase. Spacecraft separation from the *Centaur* was satisfactorily accomplished at 14:58:38 GMT. The flight control system nulled the small rotational rates imparted by the separation springs and initiated the roll-yaw sun acquisition sequence. After a minus roll of 100.5 deg and a plus yaw of 86 deg, sun acquisition and lock-on were satisfactorily completed at 15:03:20.

At 15:53:39, a positive roll maneuver was initiated to generate a star verification map to identify Canopus for the purpose of establishing the spacecraft roll axis orientation in space. The upper gate limit on the Canopus sensor was raised prior to launch to 1.5 × Canopus intensity to compensate for an observed sensor operating degradation of ½% per hr under high sun test simulation. During mapping, Canopus intensity was outside the upper gate limit. Without a usable Canopus lock signal, manual lock command was used to lock on Canopus. Calculation of the actual spacecraft roll rate was made from the star map using the time between successive passes on stars Alderamin, Merak, Regulus, Alpha Hydrae, and Naos. Roll rate was determined to be 720.4 sec for a 360-deg angle or 0.4997 deg/sec as compared to a system design of 720 sec for a 360-deg angle or 0.5-deg/sec roll rate.

c. Midcourse maneuver and velocity correction. Pointing of the spacecraft for a velocity correction was accomplished at 06:36:45 (Day 151) by a −86.5-deg roll and a −57.99-deg yaw. The vernier thrusting period was set by command for 20.8 sec. With a controlled acceleration of 0.1 g, a total velocity correction of 20.4 m/sec was achieved.

Startup of the verniers resulted in exceptionally smooth transients to the spacecraft, indicating that the verniers ignited and became throttleable almost simultaneously. Peak gyro errors from vernier ignition were 0.15 deg for pitch, 0.15 deg for yaw, and 0.10 deg for roll.

Shutdown transients resulted in peak gyro errors of +0.6 deg for pitch, +1.9 deg for yaw and +0.7 deg roll. The reverse angles of +57.99 deg yaw and +86.5 deg roll were commanded, resulting in a sun and star reacquisition at 07:00:56.

The vernier engine transients did not result in any inertial reference loss since the gyros have a minimum of 10-deg input capability, and peak gyro error was 1.9 deg. The reverse-angle commanded maneuvers resulted in a routine system reacquisition.

d. Cruise mode. Prior to and after the midcourse maneuver, six gyro drift rate tests were performed. The drift requirement for the *Surveyor* spacecraft gyros is 1.0 deg/hr (non g-sensitive) or less. The average values of drift were: pitch, 0 deg/hr; roll, +0.2 deg/hr; yaw, +0.75 deg/hr. The average inertial dead bands were: pitch, + 0.47 deg; roll, +0.44 deg; yaw, + 0.42 deg. The average inertial limit cycle periods were: pitch, 197 sec; roll, 283 sec; yaw, 195 sec. The average optical dead bands were: pitch, +0.23 deg; roll, +0.45 deg; yaw, +0.32 deg. The average optical limit cycle periods were: pitch, 285 sec; roll, 433 sec; yaw, 190 sec.

Just prior to the terminal descent maneuver, the attitude control gas pressure and weight were 3525 psi and 3.94 lb, respectively. The predicted usage was 0.51 lb as compared with the calculated usage of 0.55 lb. This discrepancy was attributed primarily to inaccuracies of the temperature and pressure measurements. Also, the predicted usage did not account for the gas required to cage the gyros after each drift check.

e. Terminal descent. The terminal descent was characterized by exceedingly smooth system performance. A terminal maneuver of +89.5 deg in roll, +59.9 deg in yaw, and +94.1 deg in roll was executed approximately 40 min prior to main retro ignition. A delay time (from AMR signal to vernier ignition) of 7.820 sec and a vernier thrust value of 200 lb were set and verified about 35 min prior to main retro ignition.

The AMR functioned normally, giving a mark close to the predicted time. From the AGC curve (Fig. IV-38), the signal level was −56.9 dbm at main retro ignition. Since the mark sensitivity during Flight Acceptance Tests was −92.6 dbm, a margin of 35.7 db was indicated.

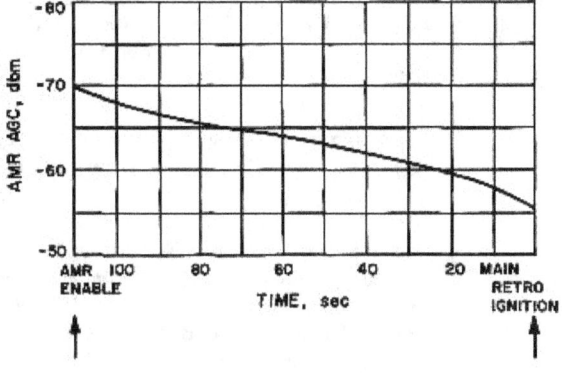

Fig. IV-38. AMR AGC level

During main retro phase, the peak deceleration was approximately 9.7 g, and the time from retro ignition to maximum vernier engine thrust was approximately 39.8 sec. During vernier ignition to case separation, attitude motion was extremely small in all three axes. Peak pitch and yaw attitude motion, as deduced from gyro error telemetry data, occurred at retro ignition and amounted to −1.0 deg in yaw, less than +0.3 deg in pitch, and +0.2 deg in roll. Following ignition, static attitude error was on the order of +0.1 deg in yaw (less than +0.3 deg required) and virtually zero in pitch. Roll attitude error remained less than +0.5 deg throughout the retro phase (less than +1.0 deg required).

Following retro ignition, at which time Vernier Engine No. 1 reached a peak thrust of approximately 90 lb, all three vernier engines settled almost precisely at their midthrust conditions. This indicated that the retro disturbance torque due to thrust vector to CG offset was virtually nonexistent. At about 10 sec prior to 3.5-g burnout occurrence, 10 ft-lb of control torque was produced by differential thrusting of vernier engines 1 and 3. Assuming 9,000 lb of retro thrust, this implies a thrust vector to CG offset of approximately 0.013 in., whereas a total of 0.18 in. of offset is allowed. The small offset attests to the excellent accuracy achieved in prelaunch retro alignment and small CG shift during retro burn.

During the high thrust phase prior to retro motor case separation, a net vernier thrust of approximately 270 lb was generated (273 ±20 lb predicted). No attitude disturbance was noted at case ejection. The estimated roll disturbance torque on the spacecraft was extremely small during both midcourse and terminal maneuvers. The principal causes of vehicle roll disturbances were vernier engine bracket bending (predicted) induced by operation of the vernier engines and thrust vector excursions of the three vernier engines. Based on telemetry data, the actuator angle was less than 1.1 deg for all but 2 sec of the mission.

RADVS lockup occurred during retro fire, and correct readings of velocities and range were obtained as soon as values were below telemetry saturation. However, beam 3 broke lock at 113.0 sec before touchdown. Relock was obtained 2 sec later. The loss of lock was probably caused by an intersection of beam 3 by the main retro casing after ejection. There was a loss of doppler and *range reliable* signals during this period as was expected for this condition. Since the *reliable* signals were pre-

viously available for greater than 1 sec, the *conditional reliable* mode was not utilized and the vernier engines went to low thrust (inertial hold) during the period.

During the 0.9-g descent phase following case separation, the RADVS satisfactorily controlled the spacecraft thrust axis with reference to the velocity direction. About 22 sec of RADVS control was experienced prior to acquisition of the descent command segments.

A profile of range vs velocity is given in Figs. IV-39, 40. The straight line corresponds to the segments mechanized in the flight control system aboard the spacecraft. The dashed line is the predicted flight path and the x marks are the actual spacecraft data points as obtained from the telemetry. The break in the curve in Fig. IV-39 between 800 and 1000 ft is due to telemetry lag at the scale change and was predicted.

The lunar reflectivity characteristics were investigated using terminal descent data. Reflectivity values for all four (RADVS) beams are presented in Figs. IV-41 through IV-44. It was assumed that the RADVS system and telemetry were linear, in estimating lunar reflectivity variation from expected values, and this should be an adequate estimation for accuracies within 1 or 2 db. Figures IV-41 through IV-44 show an average level of 4 to 6 db below the expected *average* lunar backscattering values. This is to be expected since the surface was much smoother than average in the region where *Surveyor I* landed. (A value of 6 db below average was allowed for design purposes.) However, the instantaneous values of signal strength vs time would show considerably greater fluctuations than that shown, since the telemetry does some averaging. Three different gain states are possible (GS-1, GS-2, or GS-3) in the RADVS system. Only data for GS-1 and GS-2 are shown in the figures for clarity of presentation. Note that the scale factor is different for GS-1 and GS-2.

The altitude at engine cutoff and the touchdown velocity were investigated. Using the time between the 14-ft mark and touchdown, the spacecraft velocity gain at touchdown was 1.5×5.3 (lunar $g \approx 5.3$ ft/sec^2) or approximately 8.0 ft/sec.

Just prior to engine cutoff (14 ft mark) the spacecraft was commanded to about 4 ft/sec. The touchdown velocity was thereby estimated at 11.97 ft/sec maximum. The altitude at vernier engine cutoff therefore appeared to be 12 ft rather than 14 ft.

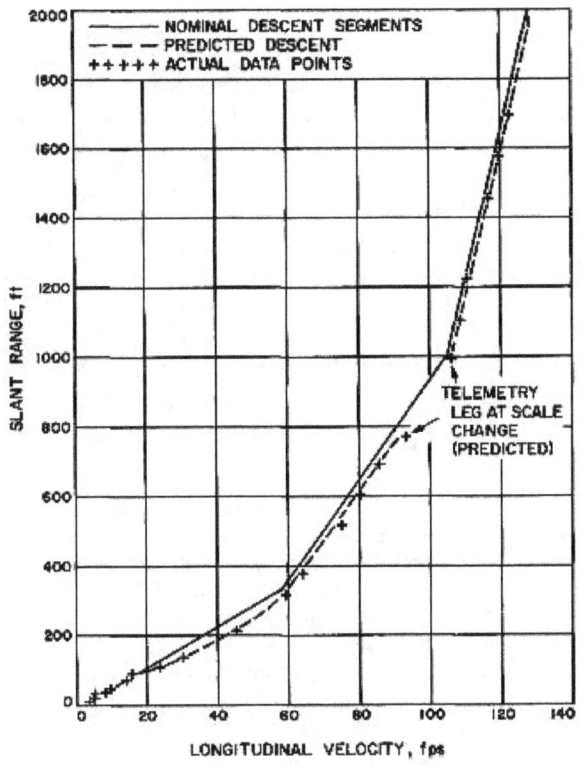

Fig. IV-39. Mission A descent profile, below 2000 ft

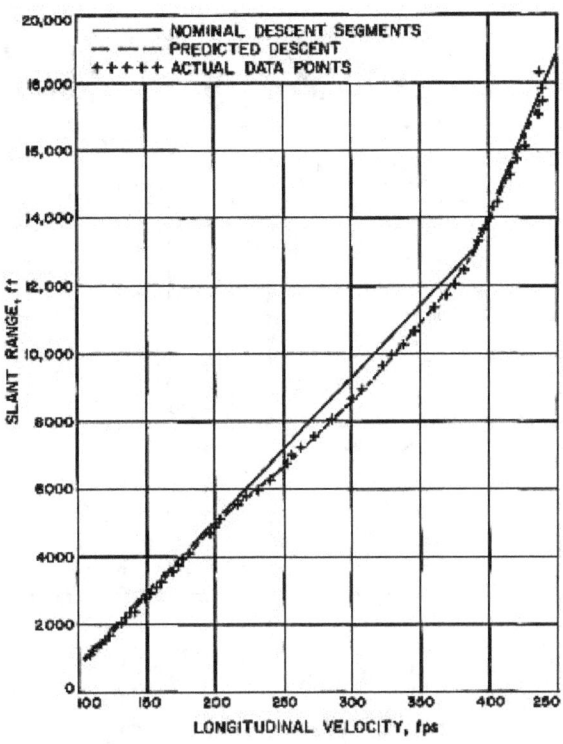

Fig. IV-40. Mission A descent profile, above 1000 ft

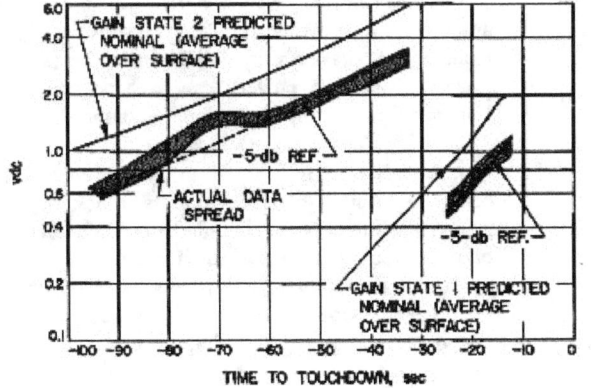

Fig. IV-41. RADVS beam-1 reflectivity signal

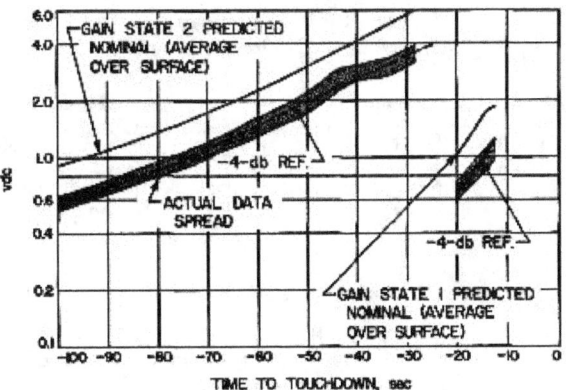

Fig. IV-42. RADVS beam-2 reflectivity signal

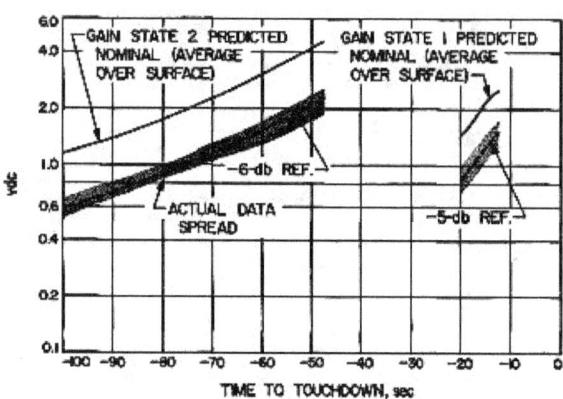

Fig. IV-43. RADVS beam-3 reflectivity signal

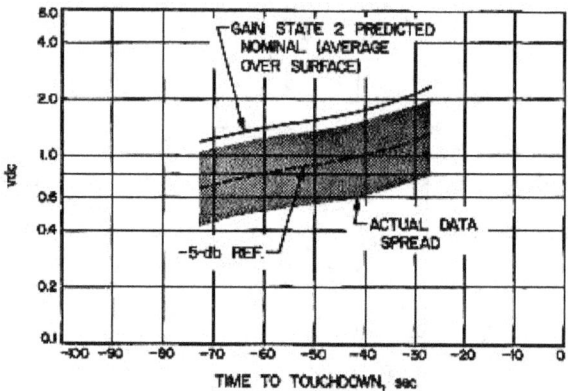

Fig. IV-44. RADVS beam-4 reflectivity signal

The slope of the surface at touchdown was within 1 deg of the local horizontal. This was determined from the postlanded gyro error shifts and the nearly simultaneous surface contact of each of the three footpads.

Gyro outputs (deg) just prior to and after touchdown are tabulated below:

Gyro	Before TD	After TD	Output change
Yaw	−0.110	+0.400	+0.510
Pitch	−0.134	−0.468	−0.334
Roll	−0.270	0.000	−0.270

The times (GMT) of individual footpad surface contact are tabulated below:

Footpad	Time (hr:min:sec)
Leg No. 1	06:17:35.656
Leg No. 2	06:17:35.656
Leg No. 3	06:17:35.670

F. Telecommunications

The spacecraft has a two-way telecommunications subsystem that provides (1) a method of telemetering information to the earth, (2) the capability of receiving and processing commands to the spacecraft, and (3) angle tracking and one- or two-way doppler for orbit determination.

The independent, identical receivers and two central command decoders are used for enhanced reliability.

Commands pass through one of the receivers to a central command decoder and then to the subsystem decoder that controls a particular subsystem of the spacecraft, such as electrical or flight control. The system provides for continuous operation of the command link in the event that either of the spacecraft receivers or either of the two central command decoders should fail.

1. Radio Subsystem

The radio subsystem utilized on the *Surveyor* spacecraft is as shown in Fig. IV-45. Dual receivers, transmitters, and antennas provide redundancy for increased reliability, though limited, by switching reliability. Each receiver is permanently connected to its corresponding antenna and transmitter. The transmitters, which are capable of operation in two different modes (100 mw low power; 10 w high power), can each be commanded to transmit through any of the three antennas.

a. Receivers. Both receivers are identical crystal-controlled double-conversion units which operate continuously (cannot be commanded off). Each unit is capable of operation in an automatic frequency control (AFC) mode or an automatic phase control (APC) mode. The receivers provide two necessary spacecraft functions: the detection and processing of commands from the ground stations for spacecraft control (AFC and APC modes), and the phase coherent spacecraft-to-earth signal required for doppler tracking (APC mode).

Highly accurate tracking of the spacecraft requires transponders that permit two-way coherent doppler shift measurements. In this mode, a transmitter is phase-locked

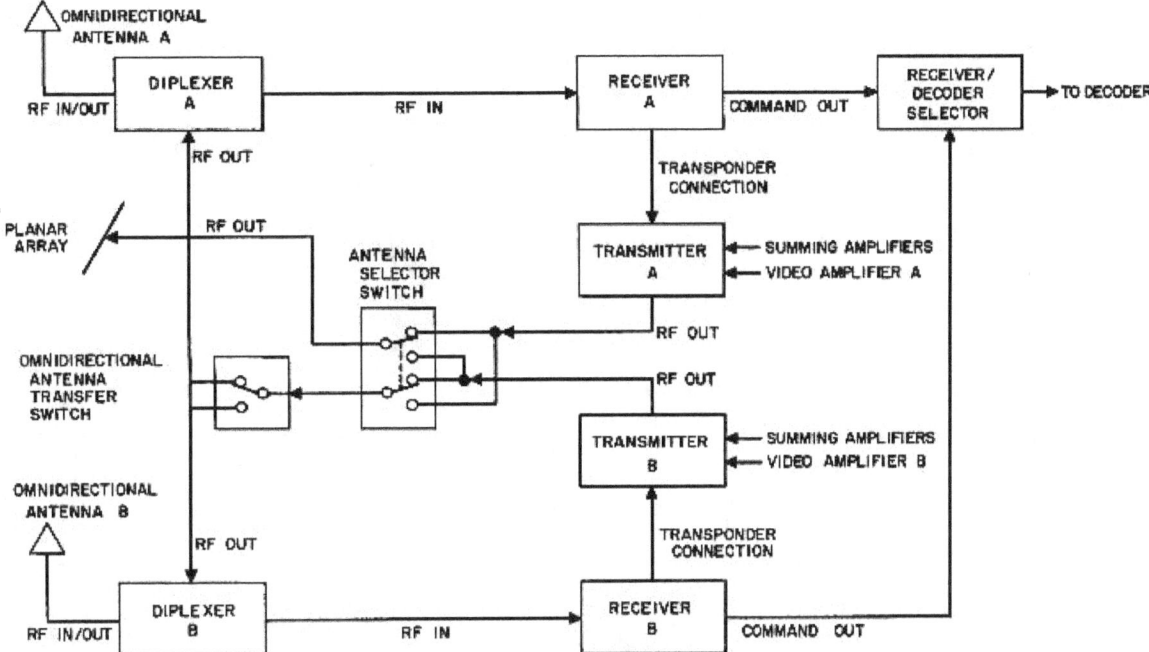

Fig. IV-45. Simplified telecommunication functional block diagram

to a receiver through transponder interconnection circuitry; the signal transmitted back to earth is phase-coherent with the received signal. On a command from earth, the spacecraft receiver operates in a phase-lock mode and tracks the frequency of the signal it receives from the DSIF station.

The raw tracking data, giving the position and velocity of the spacecraft, is actually a product of the data processing done at the DSIF stations after two-way lock is achieved.

b. Transmitters. Transmitters A and B are identical units which provide the spacecraft-to-earth link for telemetry and doppler tracking information. The transmitters are commanded on (one at a time) from the ground stations. Each unit contains two crystal-controlled oscillators (wideband for TV and scientific information, narrowband for engineering data), which can be commanded on at will, and, in addition, can operate from the receiver voltage-controlled oscillator (transponder mode) when coherent signals are required for two-way doppler tracking. Transmitters may be commanded to operate through

any one of the spacecraft antennas as desired and are both capable of providing either 100 mw or 10 w of output power.

c. Antennas. Three antennas are utilized on the *Surveyor* spacecraft. Two antennas are omnidirectional units which provide receive-transmit capability for the spacecraft. The third antenna is a high-gain (27-db) directional unit which is used for transmission of wideband information.

The radio subsystem on *Surveyor* performed nominally. Two exceptions were the "hang-up" of Omniantenna A and the high telemetry error rate experienced during the later stages of the flight. Measured signal levels on both the up and down links compared favorably with predicted levels (Figs. IV-46 through IV-48). The maximum deviation between the actual and predicted levels was 2.7 db. Frequency and power characteristics of the radio system were nominal through the mission.

During a normal mission the omniantennas are deployed by *Centaur* command just prior to separation.

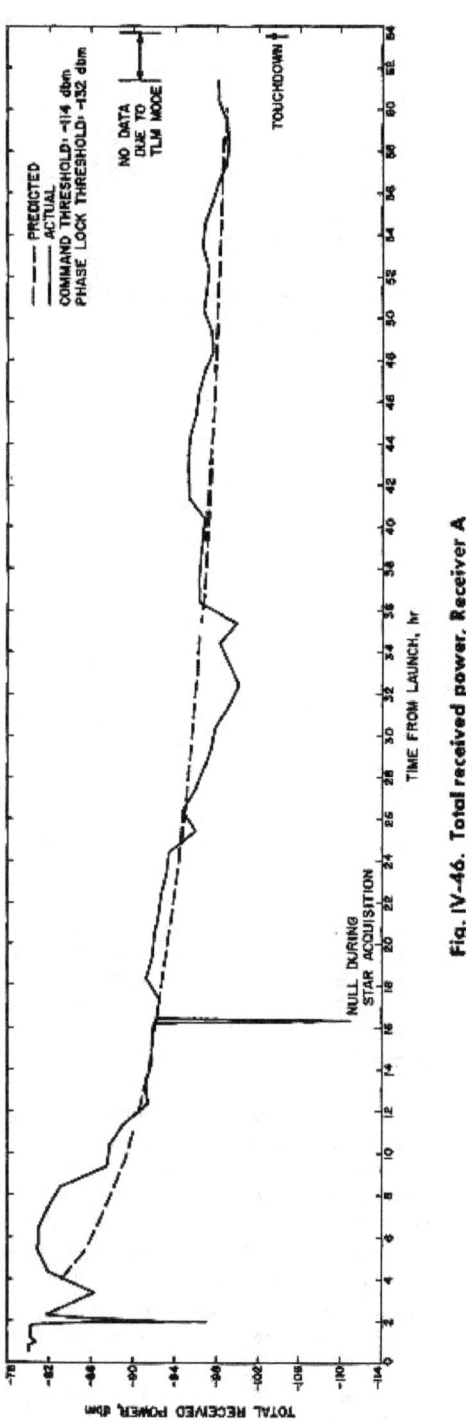

Fig. IV-46. Total received power, Receiver A

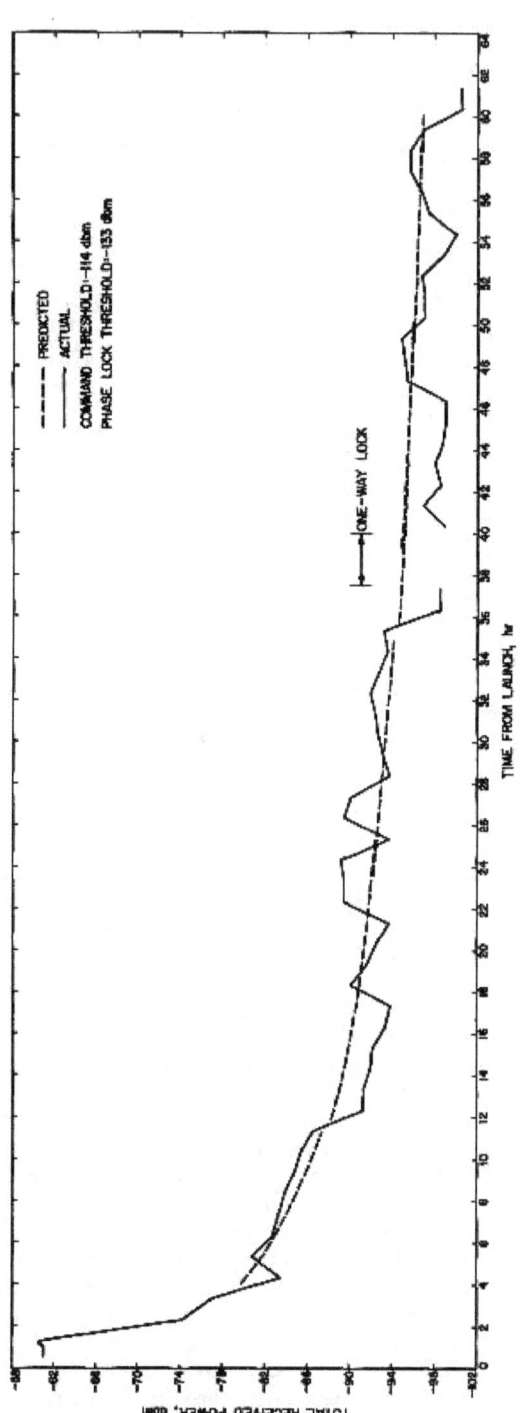

Fig. IV-47. Total received power, Receiver B

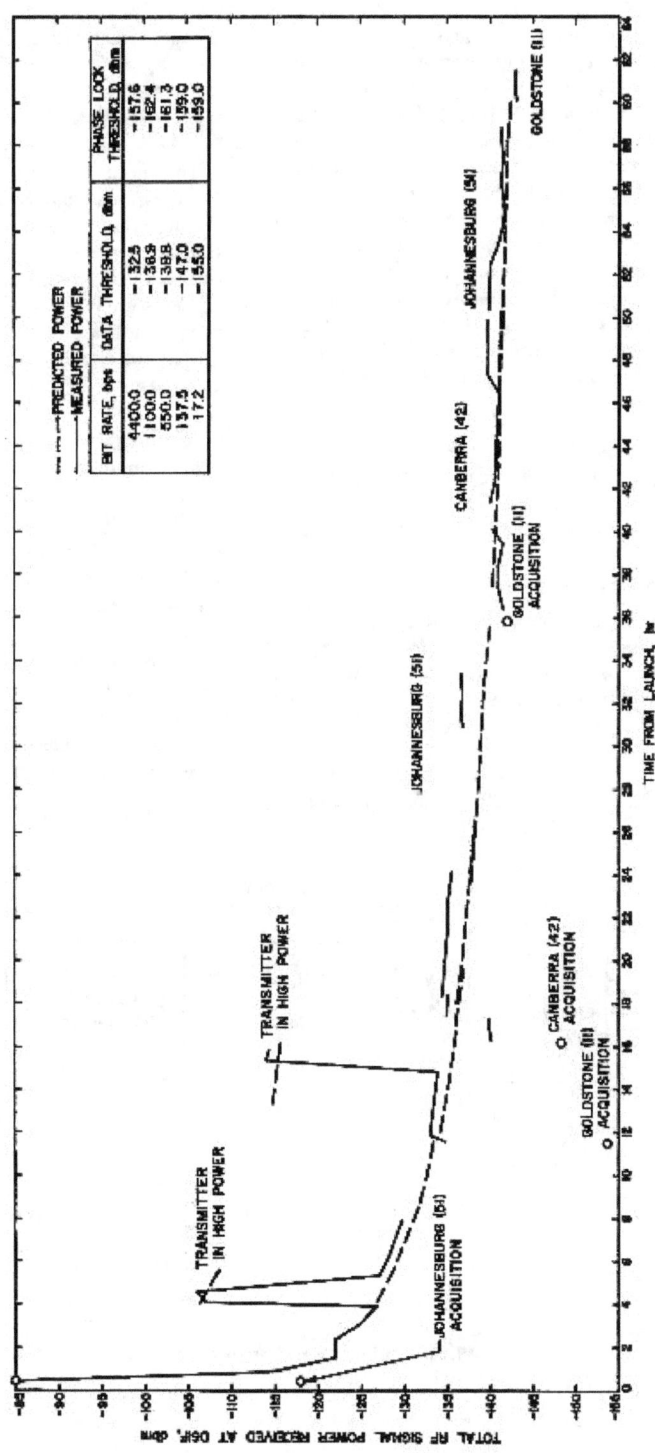

Fig. IV-48. Station received RF power

Deployment is indicated by a digital signal generated by closure of a microswitch. A normal deployment signal was received at separation for Omniantenna B; however, no signal was apparent for Omniantenna A. Examination of the receiver A AGC revealed a large null of approximately 70 deg which was not compatible with range patterns taken prior to the *Surveyor I* mission. The large unexpected nulls in the pattern indicated that proper positioning had not been achieved. During the retro sequence neither the receiver AGC nor the Omniantenna A extend telemetry was available; however, telemetry data received immediately after landing showed the antenna to be extended and locked. It is believed that the stresses encountered during terminal descent caused the antenna to free itself and deploy properly.

Bit error rates as high as 1×10^{-2} were encountered while operating in the 550 bit/sec mode during the later stages of the *Surveyor I* mission. The high error rate was expected at near lunar distance in the 550 bit/sec mode.

2. Signal Processing Subsystem

The *Surveyor* signal processing subsystem accepts, encodes, and prepares for transmission voltages, currents, and resistance changes corresponding to various spacecraft parameters such as events, voltages, temperatures, accelerations, etc.

The signal processor employs both pulse code modulation and amplitude-to-frequency modulation telemetry techniques to encode spacecraft signals for frequency- or phase-modulating the spacecraft transmitters and for recovery of these signals by the ground telemetry equipment. A simplified block diagram of the signal processor is shown in Fig. IV-49.

The input signals to the signal processor are derived from various voltage or current pickoff points within the other subsystems as well as from standard telemetry transducing devices such as strain gages, accelerometers, temperature transducers, and pressure transducers. These signals generally are conditioned to standard ranges by the originating subsystem so that a minimum amount of signal conditioning is required by the signal processor.

As illustrated in Fig. IV-49, some of the signal inputs are commutated to the input of the analog-to-digital converter while others are applied directly to subcarrier oscillators. The measurements applied directly are accelerometer and strain gage measurements which require continuous monitoring over the short intervals in which they are active.

The commutators apply the majority of telemetry input signals to the analog-to-digital converter, where they are converted to a digital word. Binary measurements such as switch closures or contents of a digital register already exist in digital form and are therefore routed around the analog-to-digital converter. In these cases,

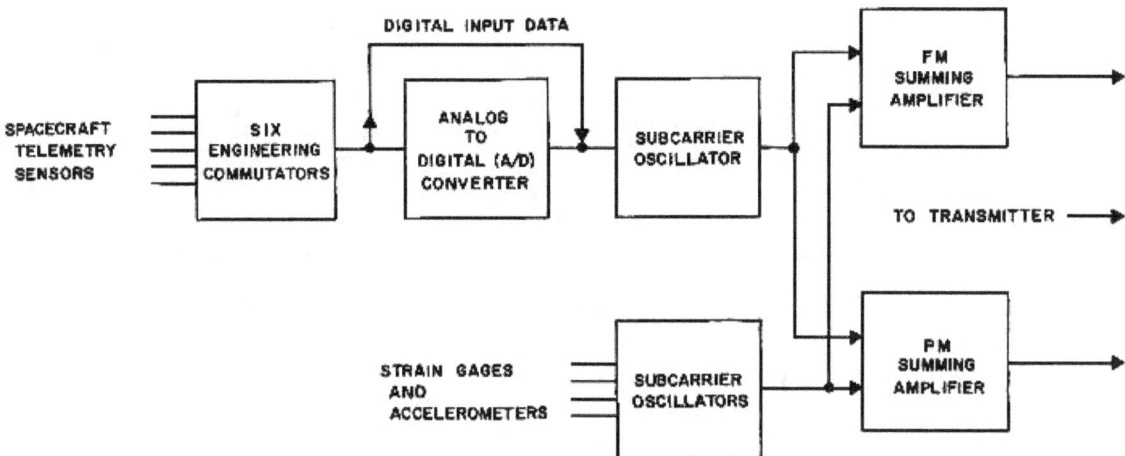

Fig. IV-49. Simplified signal processing functional block diagram

the commutator supplies an inhibit signal to the analog-to-digital converter and, by sampling, assembles the digital input information into 10-bit digital words. The commutators are comprised of transistor switches and logic circuits which select the sequence and number of switch closures. There are six engineering commutator configurations (or modes) used to satisfy the telemetry requirements for the different phases of the mission and one TV commutator configuration located in the TV auxiliary.

The analog-to-digital converter generates an 11-bit digital word for each input signal applied to it. Ten bits of the word describe the voltage level of the input signal, and one bit position is used to introduce a bit for parity-checking by the ground telemetry equipment. The analog-to-digital converter also supplies commutator advance signals to the commutators at one of six different rates. These rates enable the signal processor to supply telemetry information at 4400, 1100, 550, 137.5 and 17.2 bit/sec. The bit rates and commutator modes are changed by ground commands.

The subcarrier oscillators are voltage-controlled oscillators used to provide frequency multiplexing of the telemetry information. This technique is used to greatly increase the amount of information which is transmitted on the spacecraft carrier frequency.

The summing amplifiers sum the outputs of the subcarrier oscillators and apply the composite signal to the spacecraft transmitters. Two types of summing amplifiers are employed because of the ability of the transmitters to transmit either a phase-modulated or a frequency-modulated signal.

The signal processing subsystem employs a high degree of redundancy to insure against loss of vital spacecraft data. Two analog-to-digital converters, two independent commutators, and a wide selection of bit rate (each with the A/D converter driving a different subcarrier oscillator) provide a high reliability of the signal processing subsystem in performing its function.

During transit, 126 commands effecting changes in the signal processor were received and properly executed by the signal processor. Table IV-13 shows representative values of the telemetered signal processing parameters.

Table IV-13. Typical signal processing performance values

	Prelaunch	Flight
ESP ref. volts, v	+4.88	+4.90
EST ref. return, v	0.00	0.00
ESP unbalance current, μa	−2.96	−3.06
AESP unbalance current, μa	−1.95	−2.77
ESP full-scale current calibration, mv	91.0	90.7
ESP mid-scale current calibration, mv	50.4	50.2
ESP zero-scale current calibration, mv	10.2	10.0
AESP full-scale current calibration, mv	90.4	90.3
AESP mid-scale current calibration, mv	50.4	50.5
AESP zero-scale current calibration, mv	10.3	10.3

3. Command Decoding Subsystem

From liftoff to final touchdown on the moon the *Surveyor I* Mission required a total of 254 earth commands. Nine of these were quantitative commands (QC's) which provided the spacecraft with the quantitative information necessary for attitude and trajectory-correction maneuvers; the other 245 were direct commands (DC's) which initiated single actions such as *extend omniantennas, AMR power on, A/D clock rate 1100 bps*, etc.

These commands were received, detected, and decoded by one of the four receiver/central command decoder (CCD) combinations possible in the *Surveyor* command subsystem. The selection of the combination is accomplished by stopping the command information from modulating the uplink radio carrier for ½ sec. Once the selection is made, the link must be kept locked up continuously by either sending serial command words or unaddressable command words (referred to as fill-ins) at the maximum command rate of 2 word/sec.

The command information is formed into a 24-bit Manchester-coded digital train and is transmitted in a PCM/FM/PM modulation scheme to the spacecraft. When picked up by the spacecraft omniantennas, the radio carrier wave is stripped of the command PCM information by two series FM discriminators and a Schmitt digitizer. This digital output is then decoded by the central command decoder (CCD) for word sync, bit sync, the 5-bit address and its complement, and the 5-bit command and its complement (this latter only for DC's since the QC's contain 10 bits of information rather than 5 command bits and their complements). The CCD then compares the address with its complement and the command with its complement on a bit-by-bit basis. If the comparisons are satisfactory, the CCD then selects that one of the eight subsystem decoders (SSD) having the decoded address bits as its address, applies power to its command matrix, and then selects that one of the 32 matrix inputs having the decoded command bits as its address to issue a 20-msec pulse which initiates the desired single action.

Those DC commands that are irreversible or extremely critical are interlocked with a unique command word. Ten of the DC's and all of the QC's are in this special category. None of these commands can be initiated if the interlock command word wasn't received immediately prior to the unique command.

The QC's, besides being interlocked, are also treated somewhat differently by the command subsystem. The only differences between the DC and QC are: (1) A unique address is assigned the QC words; (2) the QC words contains 10 bits of quantitative information in place of the 5 command and 5 command complement bits. Hence, when this unique QC address is recognized, the CCD selects the flight control sensor group (FCSG) subsystem decoder (SSD) and shifts the 10 bits of quantitative information into the FCSG magnitude register. Hence, the QC quantitative bits are loaded as they are decoded.

Each picture-taking sequence after touchdown was earth-commanded. The mirror positions, the focus, etc., had to be earth-commanded. An estimated 86,000 DC's were sent to get approximately 10,000 pictures during the first lunar day. Of these thousands of commands only two were rejected by the spacecraft command system, and this was due to a problem at the ground station rather than a spacecraft command decoder failure.

G. Propulsion

The propulsion subsystem supplies thrust force during the midcourse correction and terminal descent phases of the mission. The propulsion subsystem consists of a vernier engine system and a solid-propellant main retrorocket engine. The propulsion subsystem is controlled by the flight control system through preprogrammed maneuvers, commands from earth, and maneuvers initiated by flight control sensor signals.

1. Vernier Propulsion

The vernier propulsion subsystem supplies the thrust forces for midcourse maneuver velocity vector correction, attitude control during main retrorocket engine burning, and velocity vector and attitude control during terminal descent. The vernier engine system consists of three thrust chamber assemblies and a propellant feed system. The feed system is composed of three fuel tanks, three oxidizer tanks, a high-pressure helium tank, propellant lines, and valves for system arming, operation, and deactivation.

Fuel and oxidizer are contained in six tanks of equal volume with one pair of tanks for each engine. Each tank contains a Teflon expulsion bladder to permit complete and positive expulsion and to assure propellant control under zero-g conditions. The oxidizer is nitrogen tetroxide (N_2O_4) with 10% by weight nitric oxide (NO) to depress the freezing point. The fuel is monomethyl hydrazine monohydrate (72 MMH · 28 H_2O). Fuel and oxidizer ignite hypergolically when mixed in the thrust chamber. The total usable propellant load is 178.3 lb. The arrangement of the tanks on the spaceframe is illustrated in Fig. IV-50. Propellant freezing or overheating is prevented by a combination of active and passive thermal controls, utilizing surface coatings, multilayered blankets, and electrical and solar heating. The propellant tanks are thermally isolated from a spaceframe to insure that the spacecraft structure will not function as a heat source or as a heat sink.

Propellant tank pressurization is provided by the helium tank and valve assembly (Fig. IV-51). The high-pressure helium is released to the propellant tanks by activating a squib-actuated helium release valve. A single-stage regulator maintains the propellant tank pressure at 720 psi. Helium relief valves relieve excess pressure from the propellant tanks in the event of a helium pressure regulator malfunction.

The thrust chambers (Fig. IV-52) are located near the hinge points of the three landing legs on the bottom of the main spaceframe. The moment arm of each engine is about 38 in. Engine No. 1 can be rotated ±6 deg about an axis in the spacecraft X-Y plane for spacecraft roll control. Engine 1 roll actuator is unlocked after boost. Engines Nos. 2 and 3 are not movable. The thrust of each engine (which is monitored by strain gages installed on each engine mounting bracket) can be throttled over a range of 30 to 104 lb. The specific impulse varies with engine thrust.

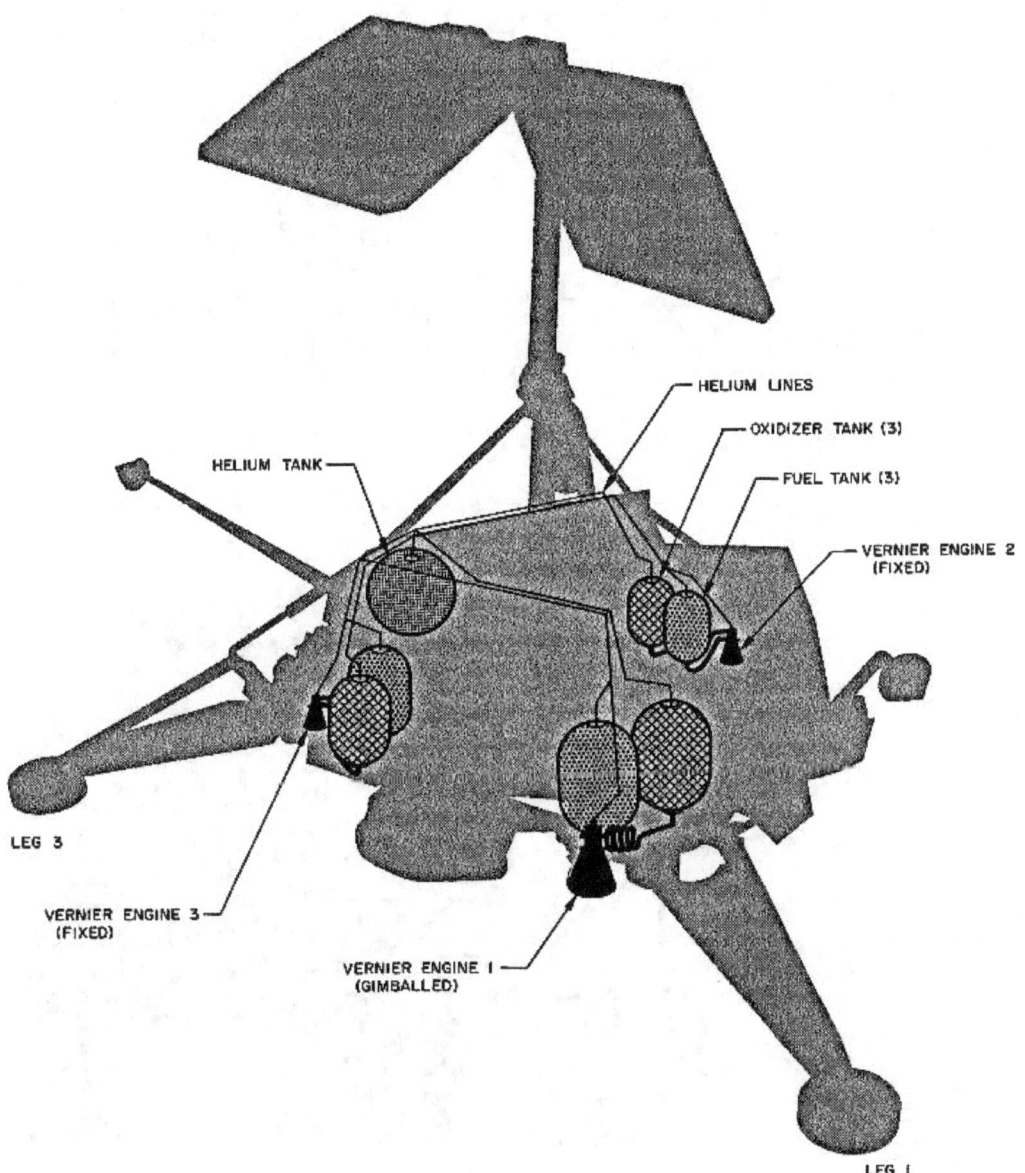

Fig. IV-50. Vernier propulsion subsystem

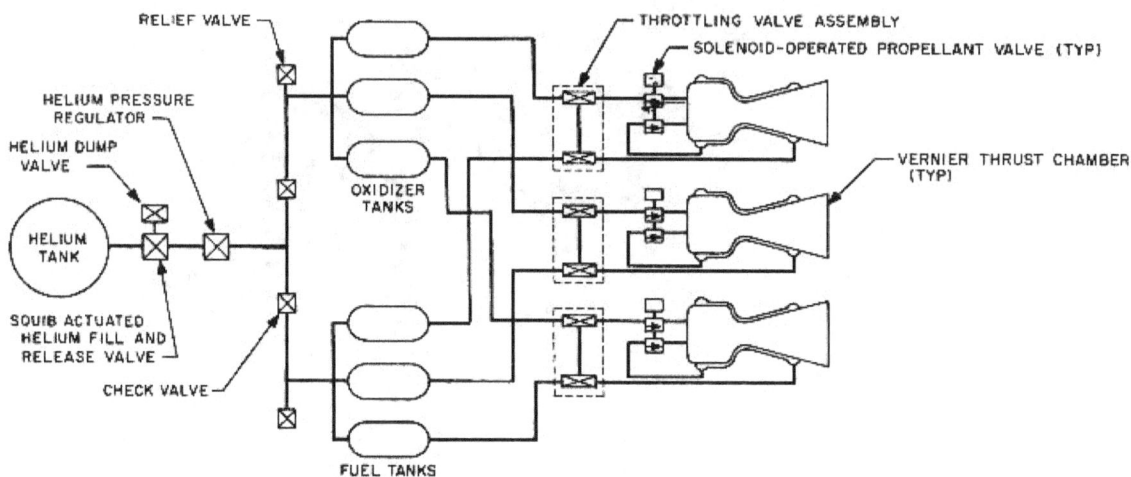

Fig. IV-51. Simplified vernier propulsion system schematic

Fig. IV-52. Vernier engine thrust chamber

SOURCE: http://heroicrelics.org/info/surveyor/surveyor-vernier.html

Prior to launch the vernier propulsion system propellant tanks are loaded with a nominal 109 lb of oxidizer and 75 lb of fuel. The propellant tanks are then pressurized to 300 psi pad pressure of helium. The high-pressure helium tank is pressurized to a nominal 5175 psi. The vernier system remains in this condition through launch until 18 min before midcourse firing, at which time a squib in the helium release valve is fired. This allows the helium regulator to pressurize the propellant tanks to their nominal working pressure of 720 psi. At midcourse the vernier system is given a command which turns on all three vernier engines to a thrust level equal to 0.1 g for a specified time depending upon the correction desired. After the midcourse maneuver the system remains in the fully pressurized state until the terminal descent sequence. For the terminal descent operation, the vernier engines are ignited 1 sec before the main retro fires.

During main retro burn, the vernier engines provide attitude control. At the end of main retro burn, the vernier engines are programmed to full thrust to facilitate the retro separation. The vernier engines are then throttled to give an optimum range/velocity profile descent. At approximately 14 ft above the lunar surface, the engines are shut down and the spacecraft free-falls to the surface.

2. Main Retrorocket

The main retrorocket, which performs the major portion of the deceleration of the spacecraft during lunar landing maneuver, is a spherical, solid-propellant unit with a partially submerged nozzle to minimize overall length (Fig. IV-53). The engine utilizes a carboxyl-terminated polyhydrocarbon composite-type propellant and conventional grain geometry.

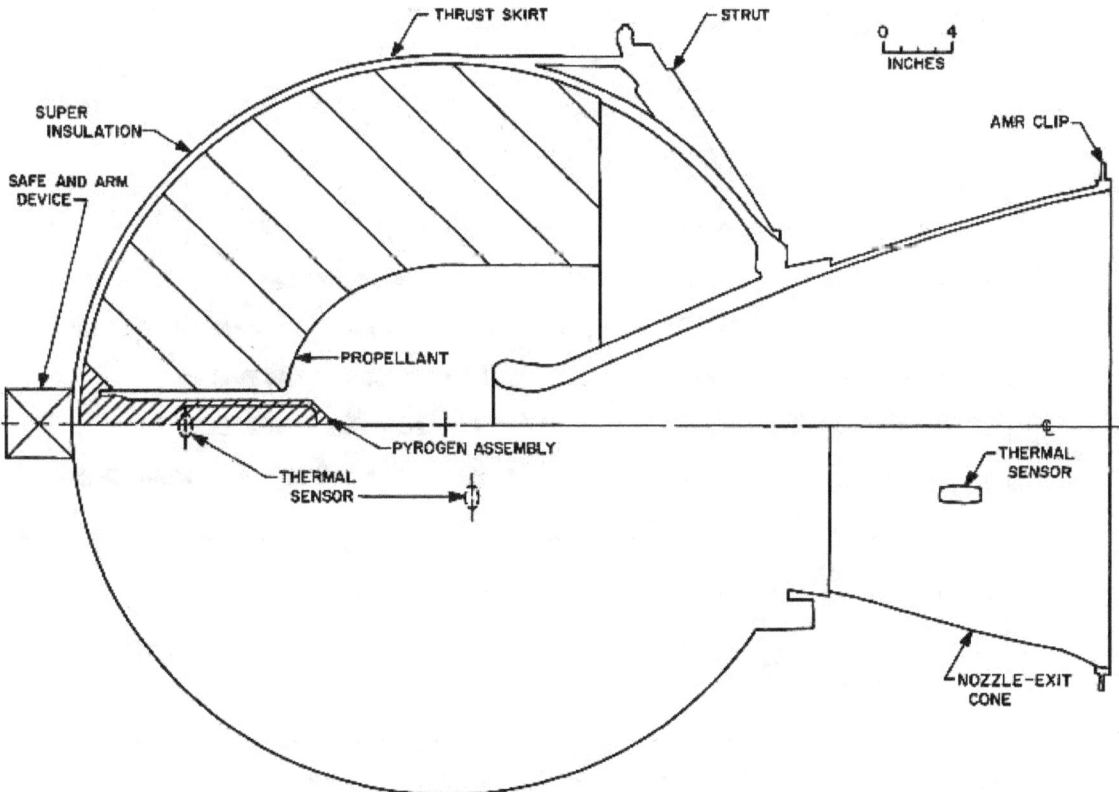

Fig. IV-53.   Retrorocket engine

The motor case is attached at three points on the main spaceframe near the landing leg hinges, with explosive nut disconnects for post-burnout ejection. Friction clips around the nozzle flange provide attachment points for the altitude marking radar (AMR). The retrorocket, including the thermal insulating blankets, weighs approximately 1390 lb. This total includes about 1244 lb of propellant. The thermal control design of the retrorocket engine is completely passive, depending on its own thermal capacity and an insulating blanket (21 layers of aluminized mylar plus a cover of aluminized Teflon). The prelaunch temperature of the unit is 70 ±5°F. At terminal maneuver, when the engine is ignited, the propellant will have cooled to a thermal gradient with a bulk average temperature of about 50 to 55°F.

The AMR normally triggers the terminal maneuver sequence. When the retro firing sequence is initiated, the retrorocket gas pressure ejects the AMR. The engine operates at a thrust level of 8,000 to 10,000 lb for approximately 39 sec at an average propellant temperature of 50°F.

3. Performance

Figure IV-54 contains a list of the major propulsion system components, their allowable temperature ranges, and their actual temperature ranges for nonmaneuver periods between launch and lunar landing.

The vernier propellant tanks were pressurized just before the midcourse maneuver. The midcourse maneuver was performed normally with a maneuver magnitude error of only 0.4%. Data taken during the midcourse maneuver are summarized in Table IV-14. The spacecraft attitude transients following midcourse ignition were less than 0.5 deg, indicating smooth and uniform ignition of the vernier engines. The spacecraft was very stable during the midcourse maneuver, indicating that the engines operated normally. Vernier engine shutdown was also smooth and uniform for all three engines as indicated by the gyro data after shutdown.

The vernier engines cooled to their steady-state temperatures within 5 hr after the midcourse correction.

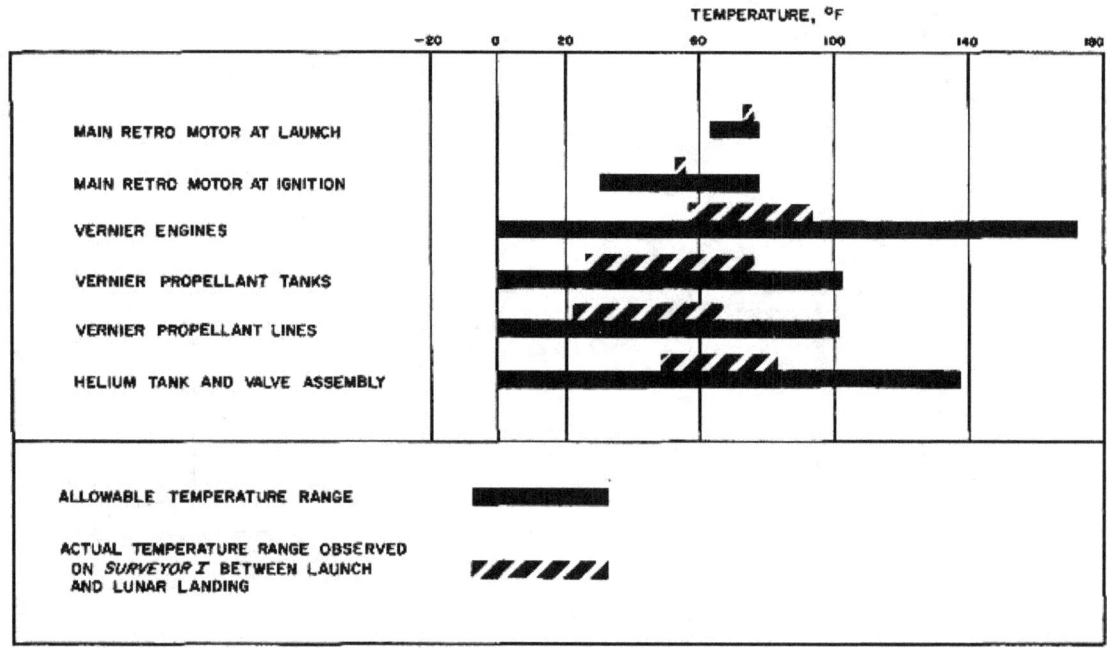

Fig. IV-54. Propulsion system temperatures, launch to landing

Helium pressurization gas consumption and regulator operation during the midcourse maneuver followed the predicted values during the midcourse maneuver as

Table IV-14. Midcourse maneuver data

Parameter	Actual value	Predicted value
Total vernier thrust, lb	219.5	219.5
Spacecraft acceleration, ft/sec²	3.21	3.21
Maneuver duration, sec	20.8	20.8
Propellant usage, lb	16.5	16.5

shown on Fig. IV-55. Propulsion system temperatures continued normal during the post-midcourse coast period as indicated in Fig. IV-54. Three of the vernier propellant tanks are equipped with thermostatically controlled heaters, but heating was not required since the temperatures remained above the thermostat settings. The temperature of the main retro propellant grain was estimated to be 54°F just before the terminal maneuver. This was within 1°F of the prelaunch prediction of 55°F at retro ignition time.

The attitude changes prior to the terminal maneuver were accomplished smoothly. The AMR was enabled

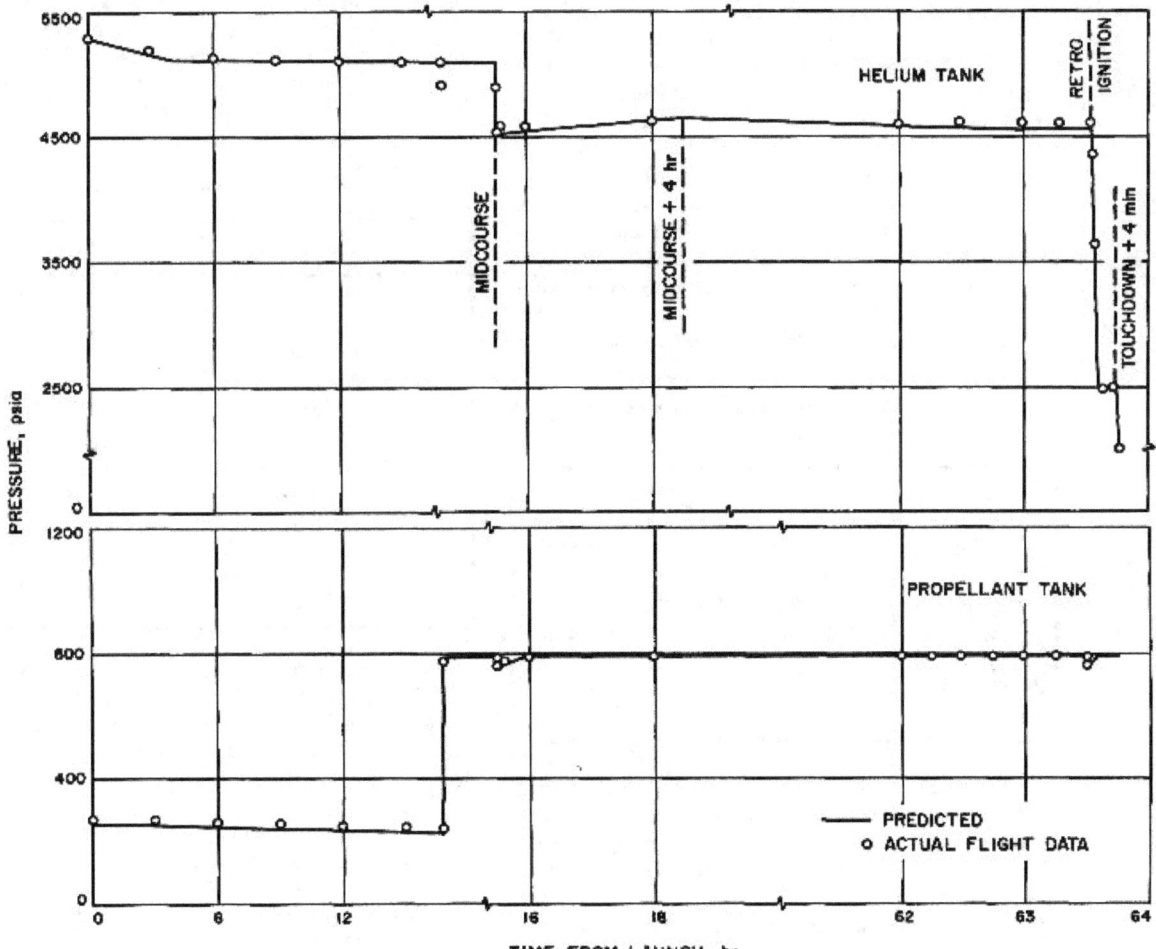

Fig. IV-55. Propellant and helium tank(s) pressure vs time

and was triggered in a normal manner to start the terminal maneuver sequence. Vernier engine ignition occurred smoothly with very small attitude transients of the vehicle. Retro ignition occurred 1 sec later as programmed by the spacecraft. On retro ignition a small attitude transient occurred (approximately 1 deg). Engine No. 1 thrust went momentarily high and then returned to approximately mid-level. Engines 2 and 3 dropped slightly and then returned to near their mid-level. It is suspected that Omniantenna A extended at this time and caused the momentary transient. During the retro maneuver, the pitch and yaw moments never exceeded 400 in.-lb, which is well within the pitch/yaw moment control capability of the spacecraft of approximately 2,000 in.-lb. This data indicated that the retro thrust vector was very well aligned with the spacecraft CG and that the retro thrust vector did not shift significantly during retro firing. The roll torques produced by the main retro were less than 25 in.-lb. This moment should be compared with the roll moment control capability of greater than 90 in.-lb. This data indicated that there was very little rotational energy in the retro exhaust. There was very little modulation of vernier engine thrust during the retro maneuver, indicating very stable operation of the vehicle. The control characteristics are summarized in Table IV-15.

Table IV.-15. Control characteristics during terminal descent

Parameter	Actual range	Allowable range
Retro phrase pitch or yaw in-lb	400 max	0-2000
Retro phase roll in-lb	25 max	0-90
Vanier engine throttling, ib	26-99	24-105
Vanier engine shutdown impulse variation, lb-sec	0.46 max	0-0.64
Pitch or yaw gyro error, drag	1° max	0-10°
Roll gyro error, drag	0.5° max	0-10°

Retro tailoff occurred smoothly, and when the inertia switch closed, the vernier engines were throttled to high thrust for case separation. The engines were operating at that time at about 99 lb each, which was the highest commanded thrust during the mission. The data indicates that the retro burning time was 38.9 sec rather than the predicted 38.5 sec. The 0.4-sec error was larger than anticipated but still within the tolerance of the spacecraft system. The data indicates that the actual total impulse of the main retro was within 0.4% of the predicted

value. Retro case separation occurred smoothly, indicating that the retro tailoff characteristics were adequate.

During the initial part of the vernier descent phase, which begins just after retro case separation, the RADVS lost lock. This was probably due to passage of the retro case through one of the RADVS beams. The RADVS locked up again and remained locked during the remainder of the vernier descent. The line segments were acquired and the spacecraft followed the programmed descent curve. RADVS noise modulation of the vernier engine thrust commands appeared nearly identical with analog computer simulations of the control system. Vernier engine cutoff appeared to be normal, with very little angular rate induced at engine shutoff. This again indicates very little difference between shutoff transients of the three engines. Terminal descent propulsion data is summarized in Table IV-16.

Table IV-16. Terminal maneuver data

Parameter	Actual value	Predicted value
Retro thrust (average), lb	9200	9370
Retro burtning time, sec	38.9	38.5
Retro total impulse variation from predicted, %	0.4	—
Total verier thrust		
Retro phase, lb	193.6	196.8
Retro case separation phase, lb	281.6	270.3
Low-accelaration descent, lb	≈105.0	108.0
High-accelaration descent, lb	≈276.0	267.0
Landing maneuver, lb	≈ 96.0	106.8
Vanier engine burning time, sec	165.0	161.0
Vanier propellant usage, lb	119.7	119.5

By integrating the thrust command data, it was determined that approximately 45 lb of the original 184 lb of vernier propellant remained in the vehicle after touchdown. The propellant consumption data is summarized in Table IV-17.

Table IV-17. Propellant consumption

Retro propellant loaded	1244.1 lb
Vernier propellant	
Total loaded	183.9
Consumed	
Midcourse movementr	16.5
Terminal movement	119.7
Unusable	———
Remaining usable	45.5 lb

The propulsion temperatures were all normal just after landing. The helium tank vent was commanded and venting occurred normally. The temperatures of Thrust Chamber Assemblies (TCA's) 2 and 3 cooled initially and then rose to approximately 180 and 225°F, respectively, and Engine 1 cooled to approximately 54°F. Propellant tank temperatures began to rise slowly after touchdown. The temperature range to which the vernier propulsion system components were exposed are shown in Fig. IV-56 for the period between landing and the last measurements taken during the first lunar night.

The vernier and main retro propulsion systems performed well within the spacecraft system requirements during Mission A. During lunar day and into the lunar night, the vernier propulsion system remained apparently leak-tight over temperature extremes well outside their normal operating range. At post-touchdown the oxidizer propellant tanks reached 110°F and the oxidizer pressure (Tank 3) rose to 835 psia. At this time the relief valve began to operate and vented the pressure down to approximately 818 psia, indicating normal relief valve operation. After approximately 90 hr past touchdown and 10 venting cycles, the pressure rose to 860 psia and then vented down to propellant vapor pressure during a 6-hr period. This venting appears to have been caused by failure of the relief valve to reseat. After the gas was vented from the oxidizer tanks, the bladders apparently expanded to completely fill the tanks and subsequently oxidizer pressure stabilized at the oxidizer vapor pressure.

H. Payload

The engineering payload provides telemetry data for evaluation of the spacecraft status and its capabilities for performing a lunar mission. Data to be provided includes vernier engine thrust, main retro engine case pressure, touchdown shock, and thermal status of critical components of the system including the structure. The engineering payload (Fig. IV-57) consists of the following:

Survey television camera

Television auxiliary

Auxiliary engineering signal processor

Auxiliary battery

Auxiliary battery control

Engineering payload instrumentation sensors

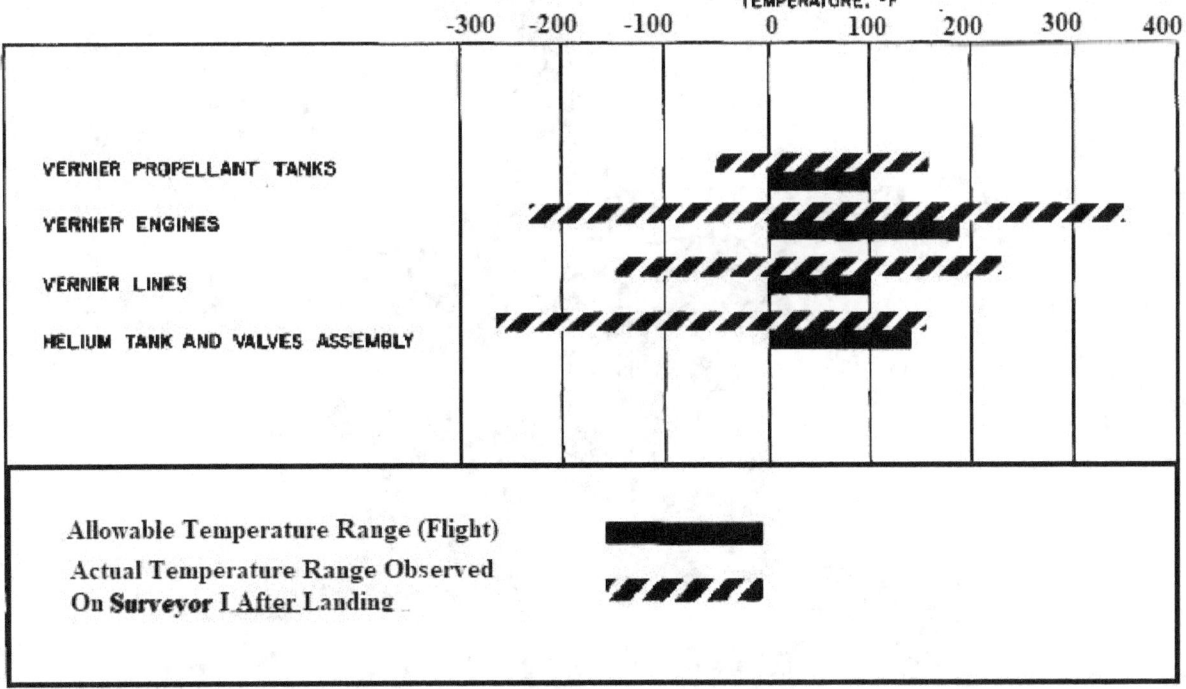

Fig. IV-56. Propulsion system temperatures, post-landing

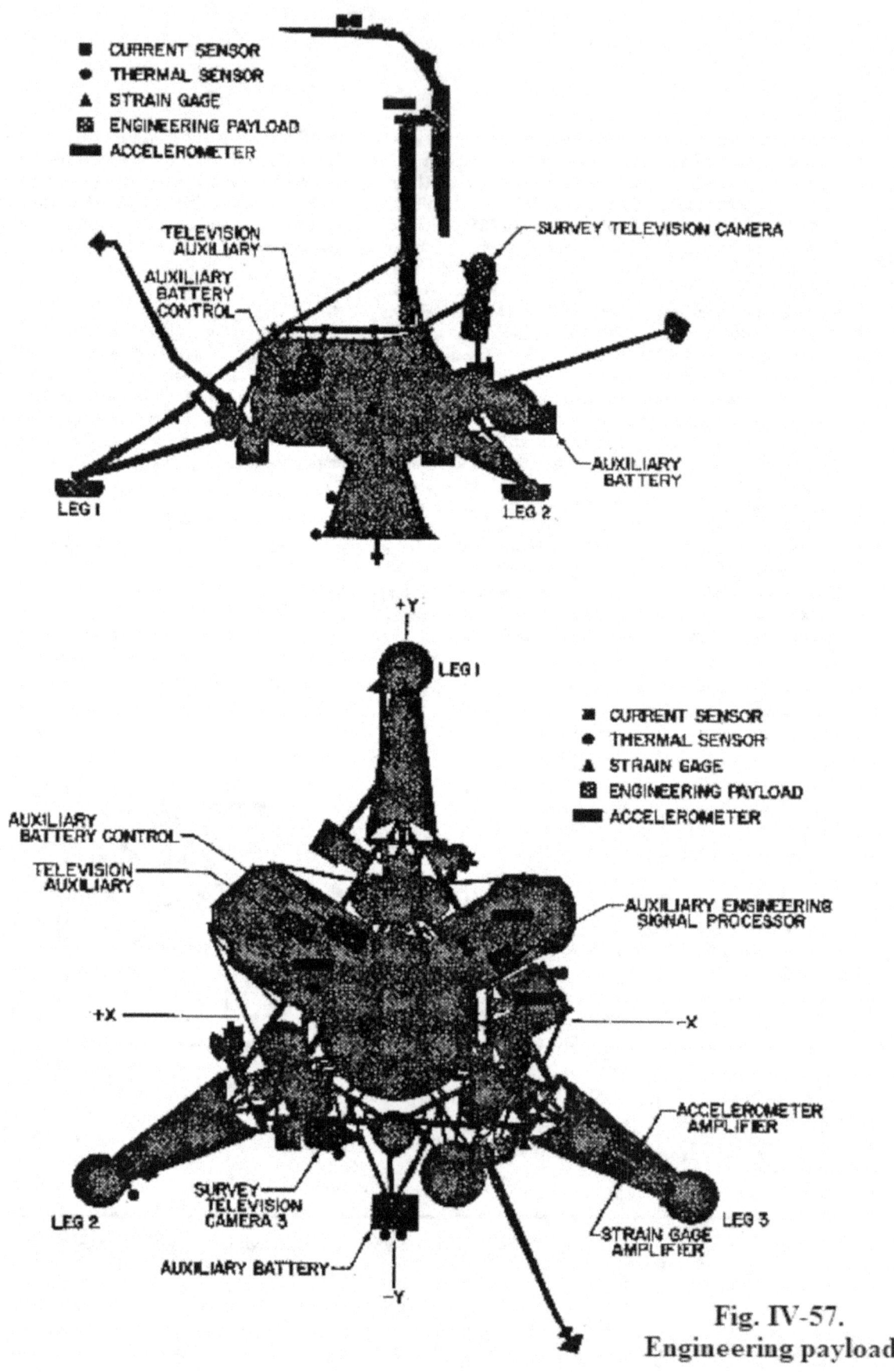

Fig. IV-57. Engineering payload

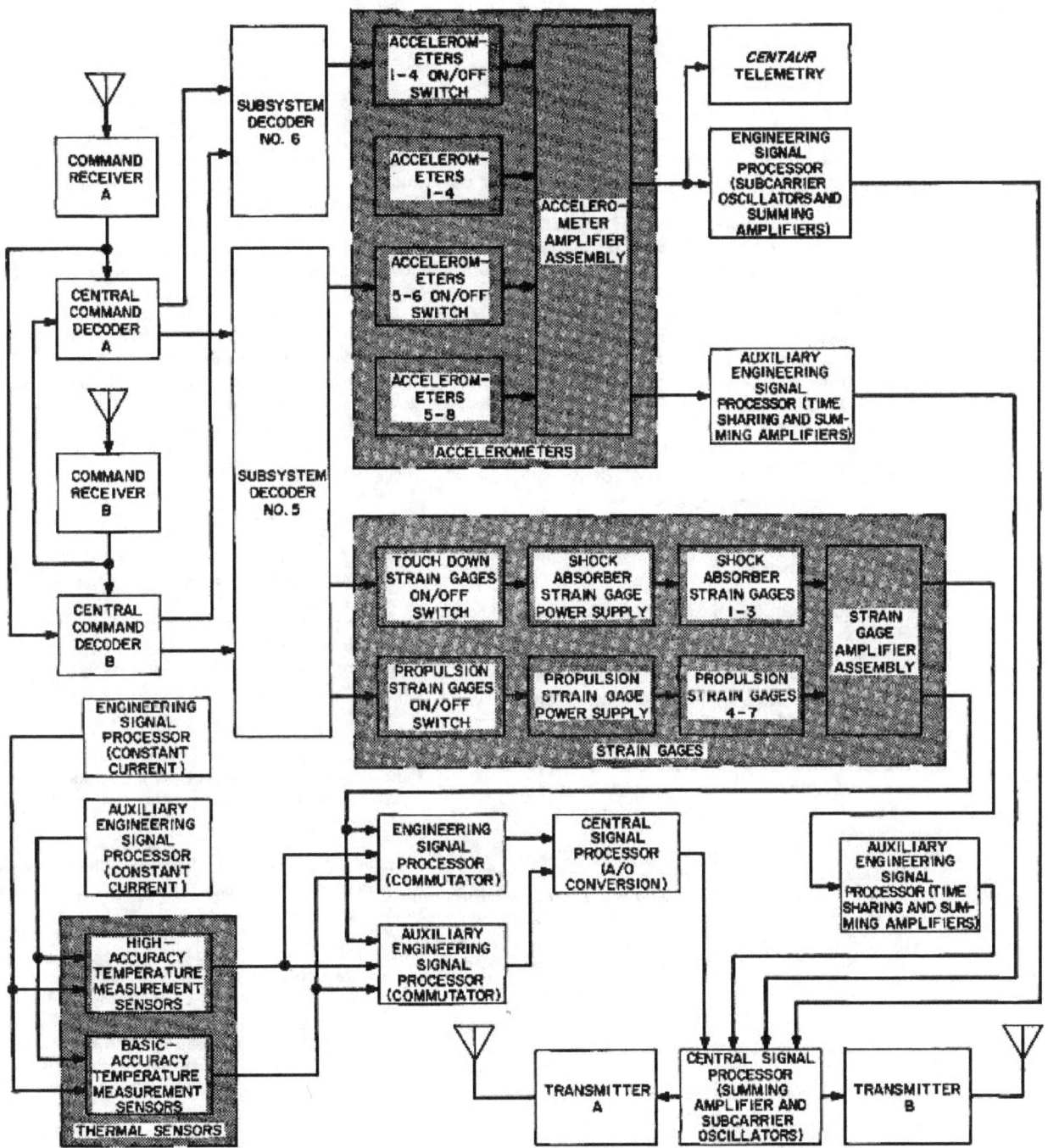

Fig. IV-58. Simplified engineering payload sensors functional block diagram

The survey television provides pictures of the lunar surface, portions of the spacecraft, and free space. It is used during lunar operations. The auxiliary engineering signal processor processes engineering payload data and other spacecraft data as a backup for the engineering signal processor for telemetry. The auxiliary battery provides a backup power source for the main battery and solar panel, and the auxiliary battery control provides for automatic and command controlled application of the auxiliary battery. The engineering payload instrumentation provides additional data on the performance status of the spacecraft and its response to the environment, beyond the capacity of the basic bus engineering instrumentation. The engineering payload instrumentation sensors (Fig. IV-58) consist of:

11 resistance thermal sensors

7 strain gages (with 1 strain-gage-amplifier assembly)

4 accelerometers

A strain gage is mounted on each of the three shock absorbers to monitor shock absorber loads during lunar touchdown. A strain gage is also mounted on each of the three vernier engine brackets to provide correlation as to vernier engine response to flight control system commands and yield a gross indication of engine thrust via the strain induced by engine operation. A strain gage on the main retrorocket engine case verifies that the engine has ignited and provides a gross indication of engine performance.

The four accelerometers are located in Compartments A and B, one on the A/SPP mast and one on the RADVS velocity-sensing antenna. Full-scale range of the accelerometer-amplifier system is ±15 g (high-gain output). The accelerometers indicate vehicle and component accelerations during all mission stages after DSIF acquisition.

Performance of the engineering payload was satisfactory and the additional data generated is elaborated in the following survey television camera section and within the spacecraft subsystem sections.

1. Television

The television subsystem is designed to obtain video photographs of the lunar surface. The subsystem consists of a downward-looking approach camera, a survey camera capable of panoramic viewing, and a television auxiliary, which serves to commutate the identification signals and provide appropriate video mixing.

The approach camera (Camera 4) is intended to provide overlapping photography of the lunar surface during the terminal phase of the trajectory and is turned on nominally 1000 nm above the lunar surface. Because of the complexity of the terminal descent phase of the mission, the approach camera was not utilized during Mission A.

The survey television camera (Camera 3) shown in Fig. IV-59 provides images of the lunar surface over a 360-deg panorama. Each picture, or frame, is imaged through an optical system onto a vidicon image sensor whose electron beam scans a photoconductive surface, thus producing an electrical output proportional to the conductivity changes resulting from the varying receipt of photons from the object space. The camera is designed to accommodate scene luminance levels from approximately 0.008 ft-lamberts to 2600 ft-lamberts, employing both electromechanical mode changes and iris control. Frame-by-frame coverage of the lunar surface provides viewing 360 deg in azimuth and from +40 deg above the plane normal to the camera Z axis to −60 deg below this same plane. Camera operation is totally dependent upon receipt of the proper command structure from earth. Commandable operation allows each frame to be generated by shutter sequencing preceded by appropriate lens settings and mirror azimuth and elevation positioning to obtain adjacent views of the object space. Functionally, the camera provides a resolution capability of approximately 1mm at 4 meters and can focus from 1.23 meters to infinity.

Figure IV-60 depicts the functional block diagram of the survey camera. Commands for the camera are processed by the central command decoder, with further processing by the TV subsystem decoder within the TV auxiliary. Identification signals, in analog form, from the camera are commutated by the television auxiliary, with analog-to-digital conversion being performed within the central signal processor. The ID data in PCM form is mixed in proper time relationship with the video signal in the TV auxiliary and subsequently sent to the telecommunications system for transmission to earth.

The survey camera consists of a mirror, filters, lens, shutter, vidicon, and the attendant electronic circuitry.

The mirror assembly is comprised of a 10.5 × 15 cm elliptical mirror supported at its minor axis by trunnions. This mirror is formed by a vacuum-depositing an aluminum surface on a beryllium blank, followed by a deposition of Kanogen with an overcoat of silicon monoxide.

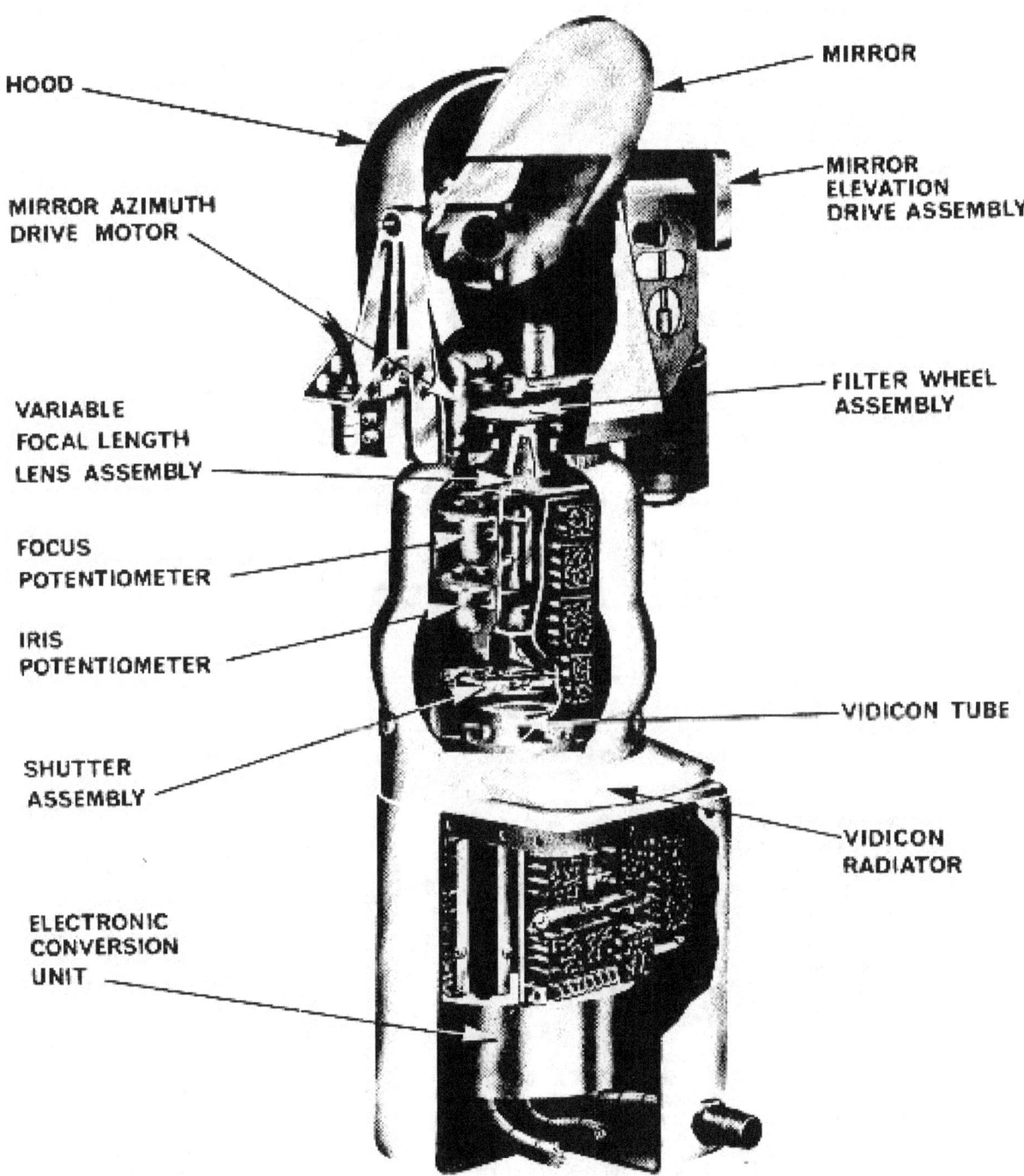

HOOD

MIRROR

MIRROR AZIMUTH DRIVE MOTOR

MIRROR ELEVATION DRIVE ASSEMBLY

VARIABLE FOCAL LENGTH LENS ASSEMBLY

FILTER WHEEL ASSEMBLY

FOCUS POTENTIOMETER

IRIS POTENTIOMETER

VIDICON TUBE

SHUTTER ASSEMBLY

VIDICON RADIATOR

ELECTRONIC CONVERSION UNIT

Fig. IV-59. Survey TV camera

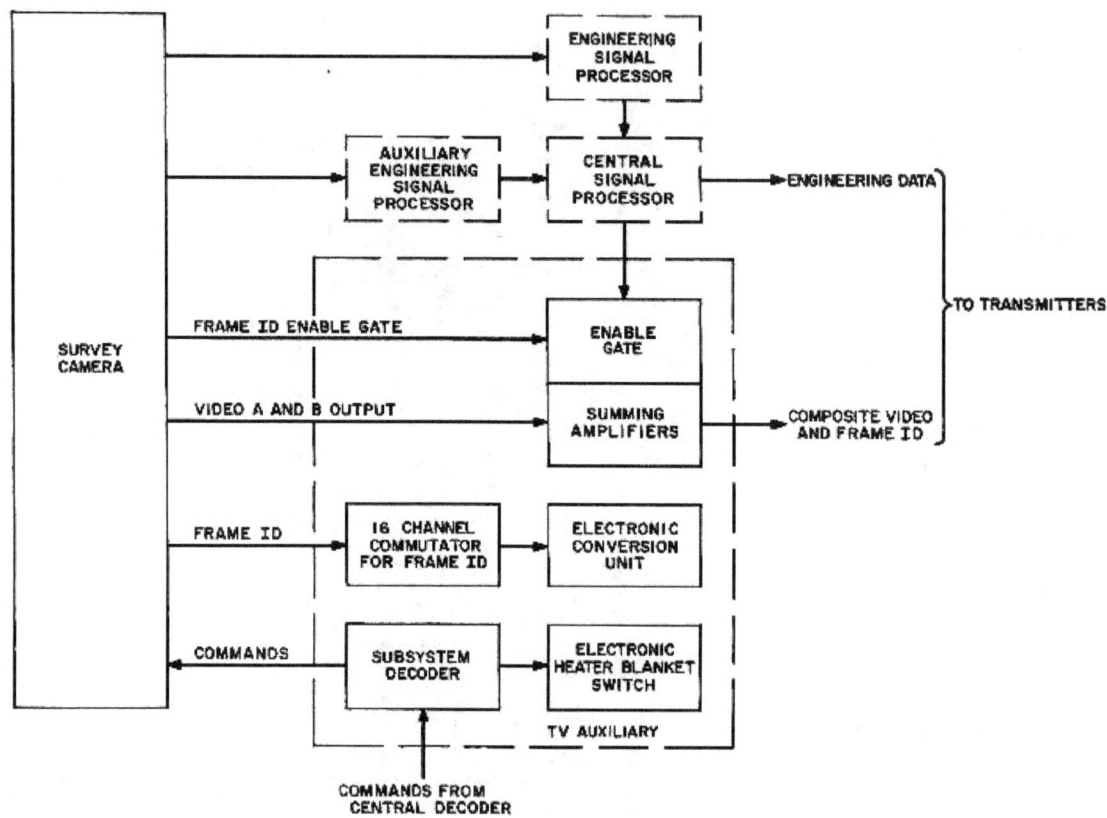

Fig. IV-60. Simplified survey TV camera functional block diagram

The mirrored surface is flat over the entire surface to less than ¼ wavelength at λ = 550 mμ and exhibits an average specular reflectivity in excess of 86%. The mirror is positioned by means of two drive mechanisms, one for azimuth and the other for elevation.

The mirror assembly contains three filters (red, green, and blue), in addition to a fourth section containing a clear element for nonmonochromatic observations. The filter characteristics are tailored such that the camera responses, including the spectral response of the image sensor, the lens, and the mirror match as nearly as possible the standard CIE tristimulus value curves (Fig. IV-61). Color photographs of any given lunar scene are reproduced on earth after three video transmissions, each with a different filter element in the field of view.

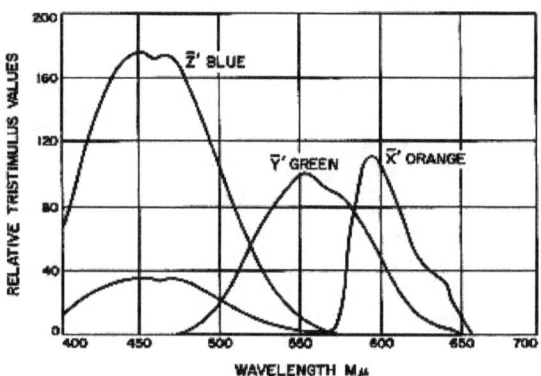

Fig. IV-61. Relative tristimulus values of the color filter elements

The optical formation of the image is performed by means of a variable-focal-length lens assembly placed between the vidicon image sensor and the mirror assembly. Each lens is capable of providing a focal length of either 100 or 25 mm, which results in an optical field of view of approximately 6.43 and 25.3 deg, respectively. Additionally, the lens assembly may vary its focus by means of a rotating focus cell from near 1.23 meters to infinity, while an adjustable iris provides effective aperture changes of from $f/4$ to $f/22$, in increments which result in an aperture area change of 0.5. While the most effective iris control is accomplished by means of command operation, a servo-type automatic iris is available to control the aperture area in proportion to the average scene luminance. As in the mirror assembly, potentiometers are geared to the iris, focal length, and focus elements to allow ground determination of these functions. A beam splitter integral to the lens assembly provides the necessary light sample (10% of incident light) for operation of the automatic iris.

Two modes of operation are afforded the camera by means of a mechanical focal plane shutter located between the lens assembly and the vidicon image sensor. Upon earth command, the shutter blades are sequentially driven by rotary solenoids across an aperture in the shutter base plate, thereby allowing light energy to reach the image sensor. The time interval between the initiation of each blade determines the exposure interval, nominally 150 millisec. An additional shutter mode allows the blades to be positioned to leave the aperture open, thereby providing continuous light energy to the image sensor. This mode of operation is useful in the imaging of scenes exhibiting extremely low luminance levels, including star patterns.

The transducing process of converting light energy from the object space to an equivalent electrical signal in the image plane is accomplished by the vidicon tube. A reference mark is included in each corner of the scanned format, which provides, in the video signal, an electronic level of the scanned image. In the normal, or 600-line mode of operation, the camera provides one 600-TV-line frame every 3.6 sec. Each frame requires nominally 1 sec to be read from the vidicon. A second mode of operation provides one 200-line frame every 61.8 sec. Each frame requires 20 sec to complete the video transmission and utilizes a bandwidth of 1.2 kc in contrast to the 220 kc used for the 600-line mode. This 200-line mode is used for omnidirectional antenna transmission from the spacecraft.

A third operational mode, used for stellar observations and lunar surface observation under earthshine illumination conditions, is referred to as an integrate mode. This mode may be applied, by earth command, to either the 200- or 600-line scan mode. Scene luminances on the order of 0.008 ft-lamberts are reproduced in this mode of operation, thereby permitting photographs under earthshine conditions.

Integral to the spacecraft and within the viewing capability of the camera are two photometric/colorimetric reference charts (Fig. IV-62). These charts, one on Omniantenna B and the other on a spacecraft leg adjacent to footpad 3, are located such that the line of sight of the camera when viewing the chart is normal to the plane of the chart. Each chart is identical and contains a series of 13 grey wedges arranged circumferentially around the chart. In addition, three color wedges, whose CIE chromaticity coordinates are known, are located radially from the chart center. A series of radial lines is incorporated to provide a gross estimate of camera resolution. Finally, the chart contains a centerpost which aids in determining the solar angles after lunar landing by means of the shadow information. Each chart, prior to launch, is calibrated goniophotometrically to allow an estimation of post-landing camera dynamic range.

The survey camera incorporates a total of four heaters to maintain proper thermal control and to provide a thermal environment in which the camera components operate. The elements are designed to provide a sustaining operating temperature during the lunar night

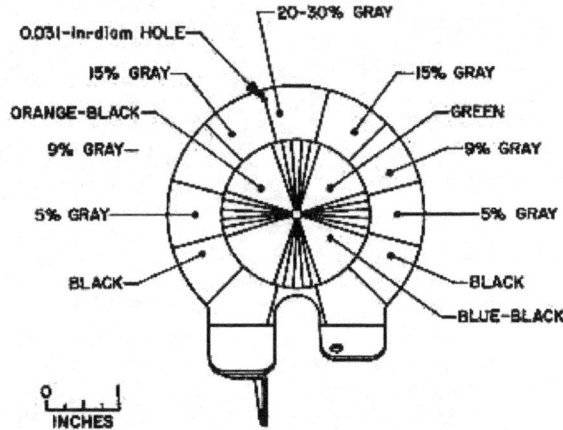

Fig. IV-62. TV photometric/colormetric reference chart

if energized. These consume 36 w of power when initiated. A temperature of −20°F must be achieved prior to camera turn on.

2. Performance

A pre-mission calibration was performed on the survey camera with the camera mounted on the spacecraft. Each calibration utilized the entire telecommunication system of the spacecraft, thereby including those factors of the modulator, transmitter, etc., which influence the overall image transfer characteristics.

The photometric and modulation transfer calibration results are shown in Figs. IV-63 through IV-68. Each of the light transfer characteristic curves is corrected for the error resulting from the spectral differences between the calibration light source and the camera components in relation to the solar spectral characteristic. These curves represent the characteristics of the camera under true lunar brightness conditions. Figure IV-68 depicts the modulation transfer of the survey camera and was obtained utilizing sine wave optical stimuli, thus enabling a determination of the true Fourier representation of the camera.

As a result of the landed roll orientation of the spacecraft, the camera was provided almost continuous shade by the solar panel/planar array, thereby allowing much longer periods of operation at high solar angles than was anticipated. The high temperature experienced during the mission was on the order of +140°F. Camera operation was discontinued following the lunar sunset, as the camera temperature dropped below −20°F point. Data indicates that the camera achieved a minimum temperature during the lunar night of −290°F. However, all components survived the lunar night, with subsequent operation showing no observable performance deteriorations as a result of the temperature extreme.

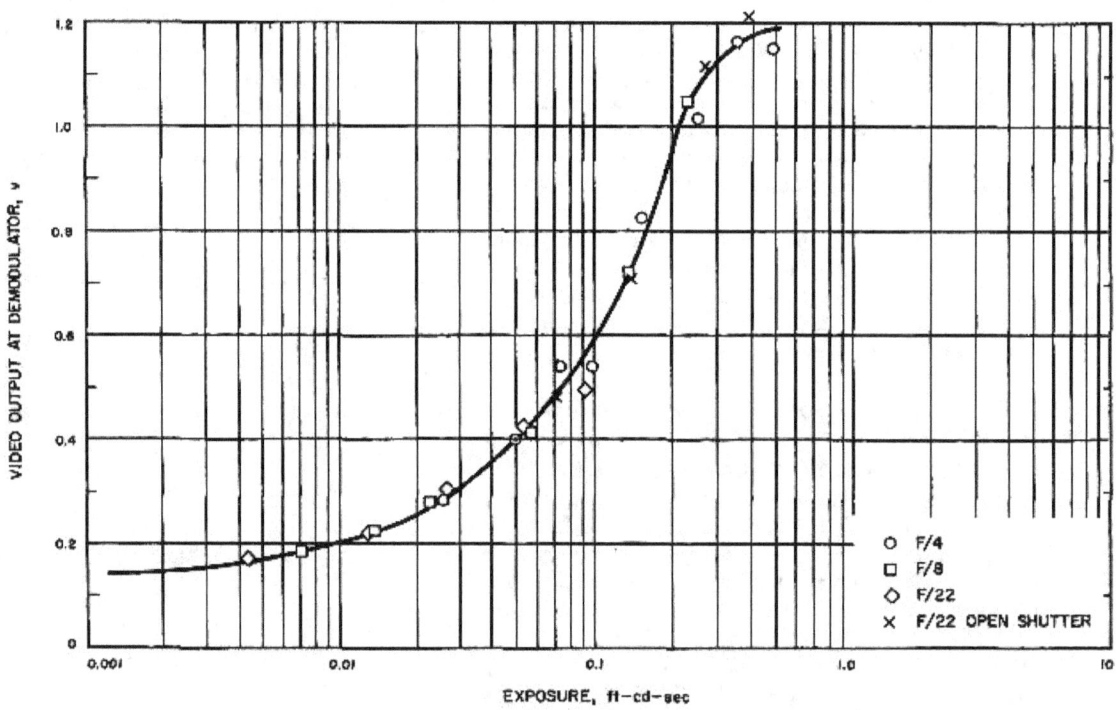

Fig. IV-63. Camera 200-line light transfer characteristic as a function of exposure

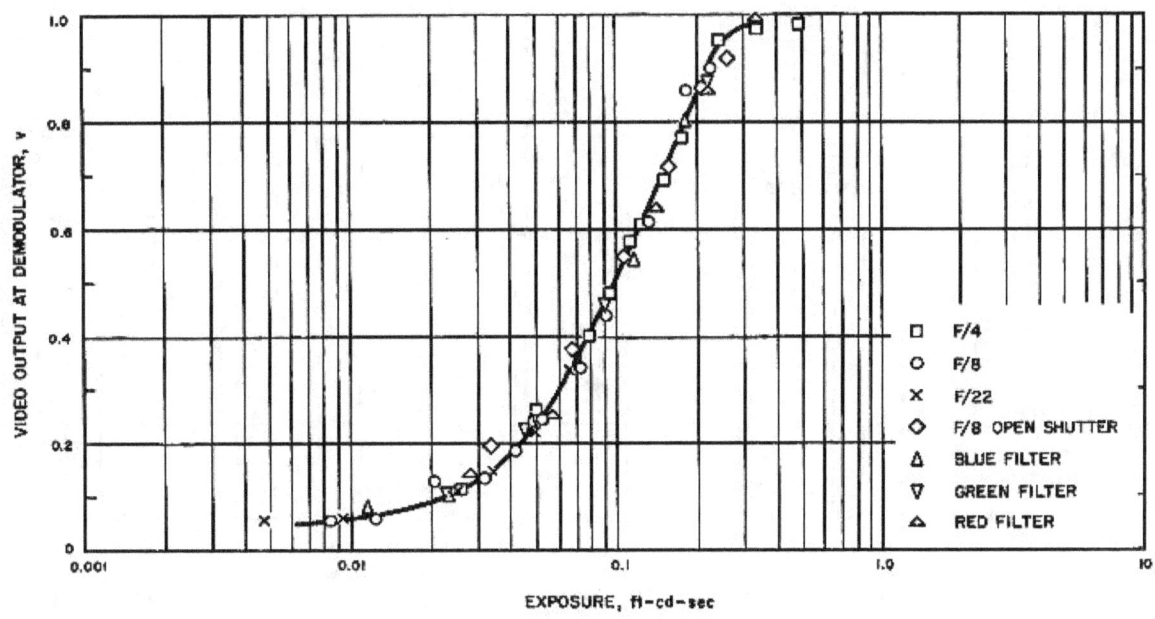

Fig. IV-64. Camera 600-line light transfer characteristic as a function of exposure

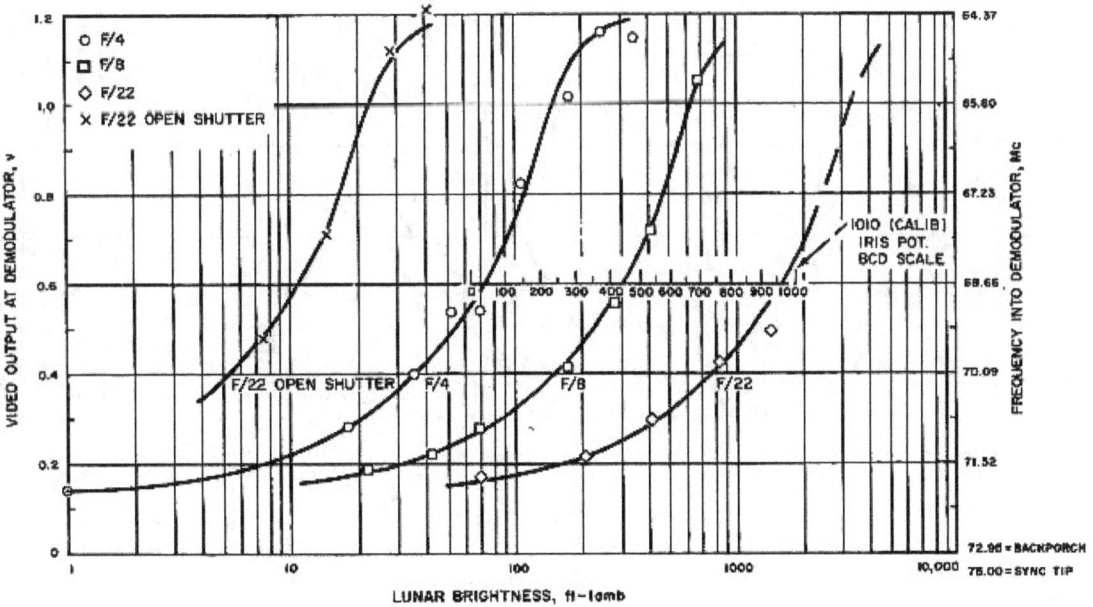

Fig. IV-65. Camera 200-line light transfer characteristic as a function of lunar brightness

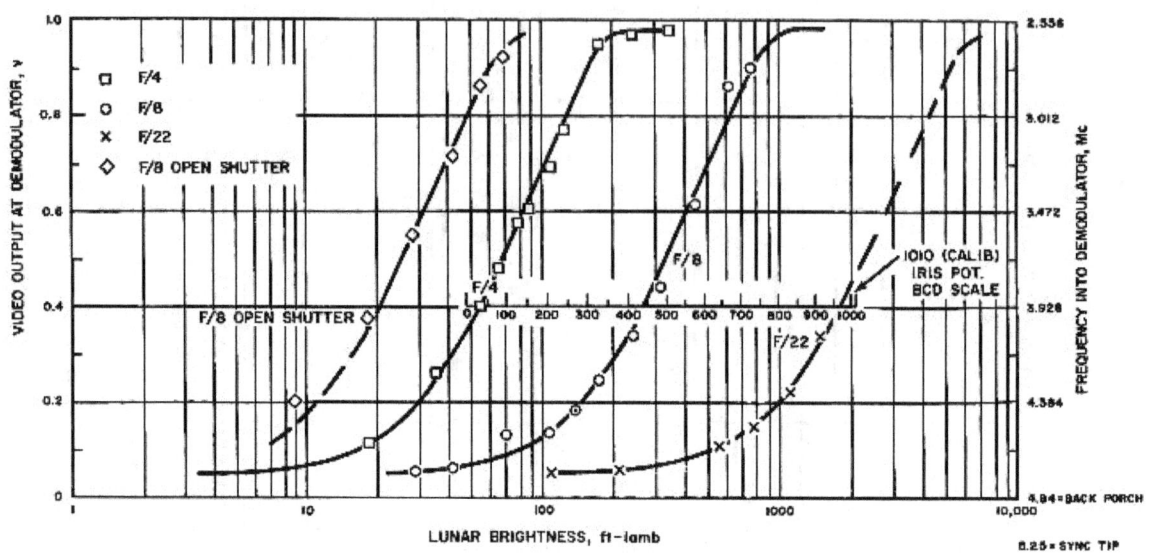

Fig. IV-66. Camera 600-line light transfer characteristic as a function of lunar brightness

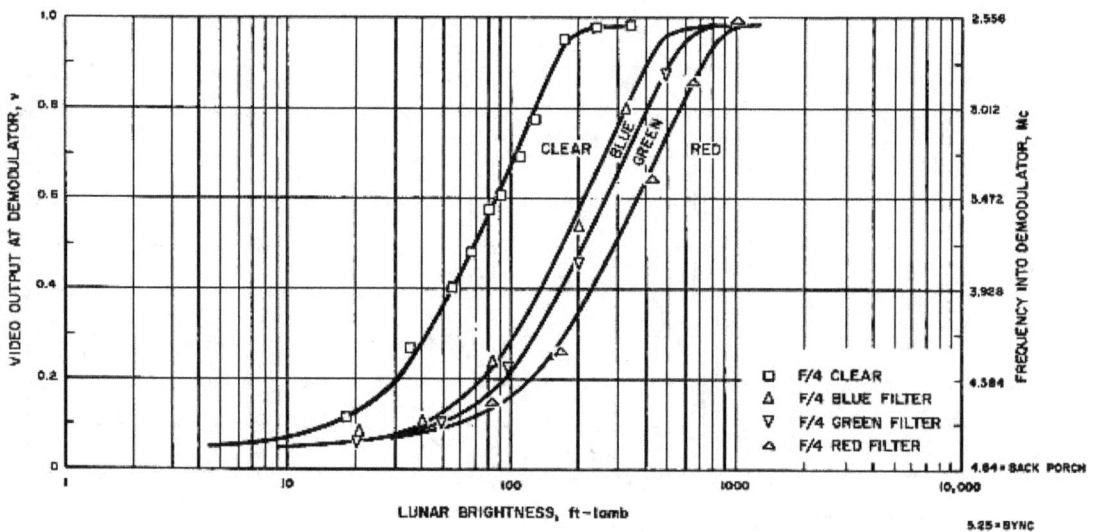

Fig. IV-67. Camera 600-line light transfer characteristic as a function of color filter position

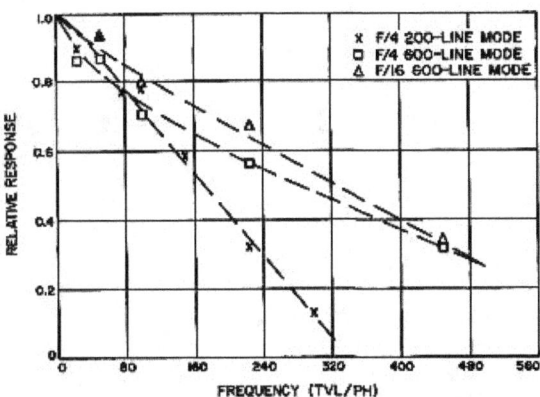

Fig. IV-68. Camera sine-wave response characteristic

3. Mission Anamolies

The camera anomalies experienced during the mission are listed below. With the exception of the first item, all anomalies were such as to be correctable with test and calibration data.

1. A mirror elevation potentiometer failed at +17 deg. The failure resulted in the loss of mirror elevation telemetry and prevented the mirror from responding to elevation commands on an intermittent basis.

2. Dirt particles were present in the camera optical system prior to launch. These particles were imaged by the camera prior to launch, thus providing a calibration on their distribution. No additional particulate material was observed in post-landing images. However, there is evidence of a change in diffuse reflectivity of the mirror, suggesting a uniform layer of fine matter on the mirror surface.

3. With a decrease in temperature of 79.7°C (143.5°F), the raster shifted left 1.5% and shifted down 1%; horizontal size decreased 4%, and vertical size decreased 6.5%.

V. TRACKING AND DATA ACQUISITION SYSTEM

The Tracking and Data Acquisition (T&DA) System for the *Surveyor* Project consists of facilities of the AFETR, GSFC, and DSN. During Mission A operations, many of the T&DA System stations encountered problems which could have rendered the conduct of space flight operations difficult if *Surveyor I* had not made a nominal flight.

A. Air Force Eastern Test Range

The AFETR performs T&DA supporting functions for *Surveyor* missions during the countdown and launch phase of flight.

The requirements for tracking and telemetry coverage of *Surveyor* missions are classified as follows in accordance with relative importance to successful mission accomplishment:

Class I requirements reflect the minimum essential needs to ensure accomplishment of first-priority flight test objectives. These are mandatory requirements which, if not met, may result in a decision not to launch.

Class II requirements define the needs to accomplish all stated flight test objectives.

Class III requirements define the ultimate in desired support. Such support should enable the range user to achieve the flight test objectives earlier in the test program.

The AFETR prelaunch configuration is shown in Table V-1. (Pretoria supported Mission A operations

Table V-1. AFETR prelaunch configuration

Station	Radar	VHF telemetry	S-band telemetry
Merritt Island	X		
Cape Kennedy	X	X	X
Patrick AFB	X		
Grand Bahama Island	X	X	X
Grand Turk	X		
Antigua	X	X	X
General Arnold	X	X	X
Coastal Crusader		X	X
Sword Knot		X	X
Ascension	X	X	X
Pretoria	X	X	X

without commitment.) A major element in the AFETR configuration is the disposition of Range Instrumentation Ships (RIS). Figure V-1 shows the planned RIS coverage for *Surveyor I* launch day. (See Fig. II-4.)

Except in the case of S-band telemetry facilities, AFETR preparations for Mission A were accomplished by routine testing of individual facilities, followed by several Operational Readiness Tests.

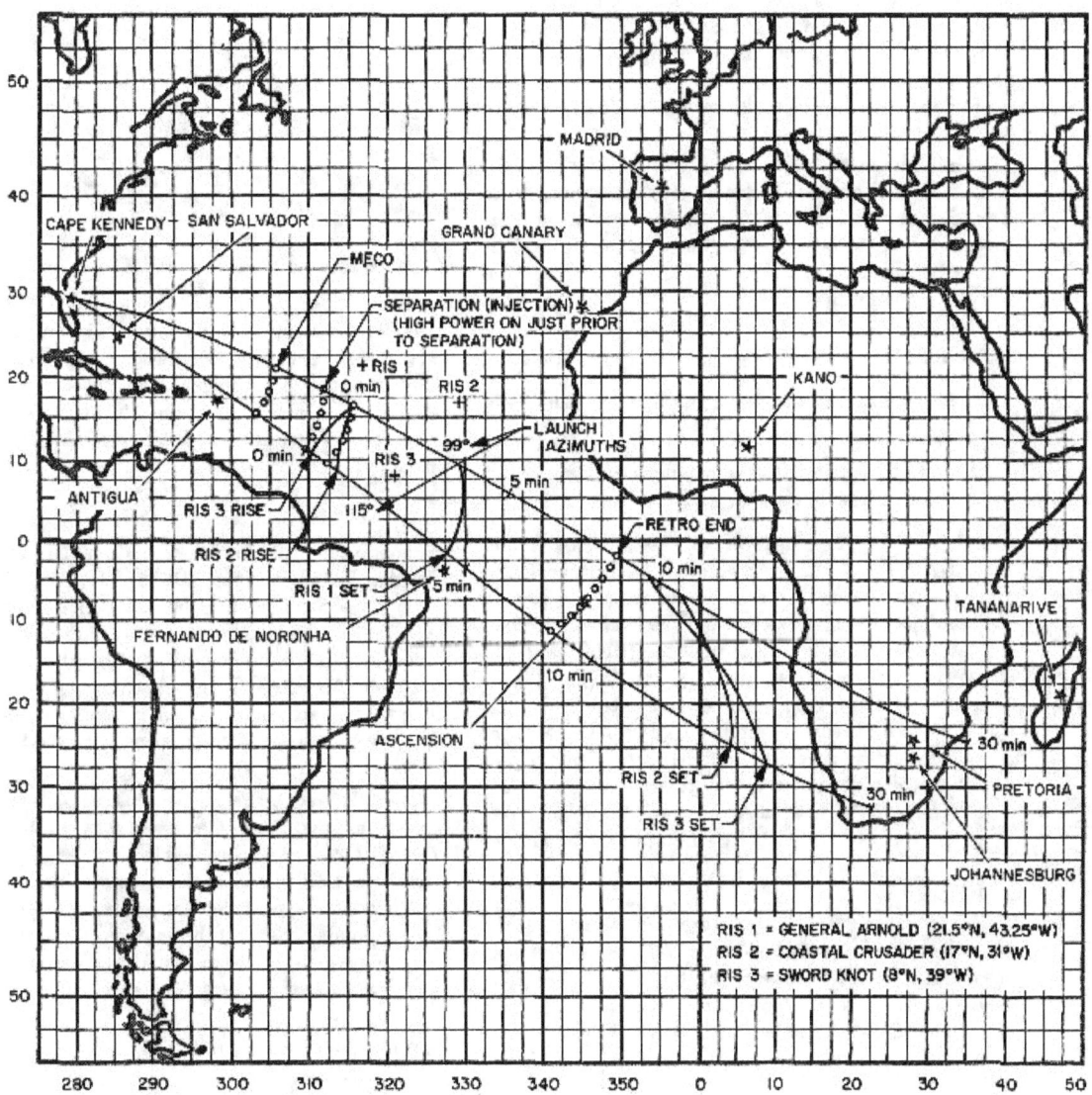

Fig. V-1. Preliminary instrumentation coverage for May 30, 1966

In general, most Class I and Class II requirements were met by AFETR for *Surveyor* Mission A.

1. Tracking (Metric) Data

The AFETR tracks the C-band beacon of the *Centaur* stage to provide metric data. This data is required during intervals of time before and after separation of the spacecraft for use in calculating the *Centaur* orbit, which can be used as a close approximation of the post-separation spacecraft orbit. The *Centaur* orbit calculations are used to provide DSN acquisition information (in-flight predicts).

Metric data needs were to be provided by the RIS *Arnold*, located downrange from Antigua. Although the *Arnold* was operating on a limited data commitment basis (because of its being under engineering cognizance), it was expected to provide data that would meet the Class I requirement.

The following anomalies occurred during countdown:

1. A noisy azimuth encoder in the Antigua radar.

2. Difficulty with slew checks of Antigua and Grand Turk radars because of subcable noise (caused by thunderstorms in the downrange area).

3. Loss of ranging capability by the *General Arnold* at T −4 min. However, all other supporting instrumentation was "go," and it was decided to launch without metric support from the *General Arnold*. At T −2 hr, a blocking oscillator circuit card had been replaced in Digital Tracker No. 1 (DT-1), after which DT-1 performed satisfactorily in pre-flight checks using a test video input. Postflight investigation revealed that the output polarity of the replacement card was incorrect, thus preventing lock-on. DT-2 had failed prior to pickup of range count and could not be repaired in time to support the flight.

The coverage intervals that were provided during the flight are shown in Figs. V-2 and V-3. As can be seen from the difference between the top bar on Fig. V-2, which shows the total expected coverage, and the bottom bar, which shows the actual interval of data, the Class I commitment was not met after *Centaur* MECO. MECO occurred at 689 sec, 6 sec later than nominal. Antigua loss of signal was expected to be at 729 sec, but actually occurred at 690 sec.

There was no commitment for metric data beyond Antigua.

The *General Arnold* radar was improperly phased at transmitter turn-on and interfered with Antigua radar operation, causing it to lose track at L +693 sec. This radar was operated in an attempt to obtain recorded angle data. The radar interference problem could have been prevented by the correct use of AFETR radar phasing procedures.

Range Safety plots were satisfactory. The plots showed the vehicle to be nominal. Grand Turk radar and Antigua radar did not meet their Range Safety commitments. The Range Safety system was satisfied, however.

2. *Atlas/Centaur* Telemetry (VHF)

Starting during the countdown, the AFETR continuously receives and records *Atlas* (229.9-mc link) telemetry until *Atlas/Centaur* separation and *Centaur* (225.7-mc link) telemetry until shortly after spacecraft separation. Thereafter, *Centaur* telemetry is recorded as station coverage permits until completion of the *Centaur* retro-maneuver.

All Class I telemetry requirements were expected to be met using the facilities of Cape Kennedy, Grand Bahama Island, Antigua, RIS *Arnold*, RIS *Coastal Crusader*, RIS *Sword Knot*, and Ascension. The coverage plan indicated some gaps in providing continuous coverage, but the requirement was only for coverage where available. The three Range Instrumentation Ships were to provide data on a limited commitment basis (since they were still under engineering cognizance).

The VHF coverage versus that actually provided is shown in Fig. V-4. Coverage was continuous from T −4500 to L +5940 sec, which is more than predicted. However, since the *Centaur* stage was not roll-attitude-stabilized, the expected coverages were based on a specified minimum db level at the antenna null. This uncertainty in antenna gain, of course, could easily result in more or less coverage than expected.

3. *Surveyor* Telemetry (S-Band)

The AFETR also receives and records *Surveyor* S-band (2295-mc) telemetry after the spacecraft transmitter high power is turned on until 15 min after injection.

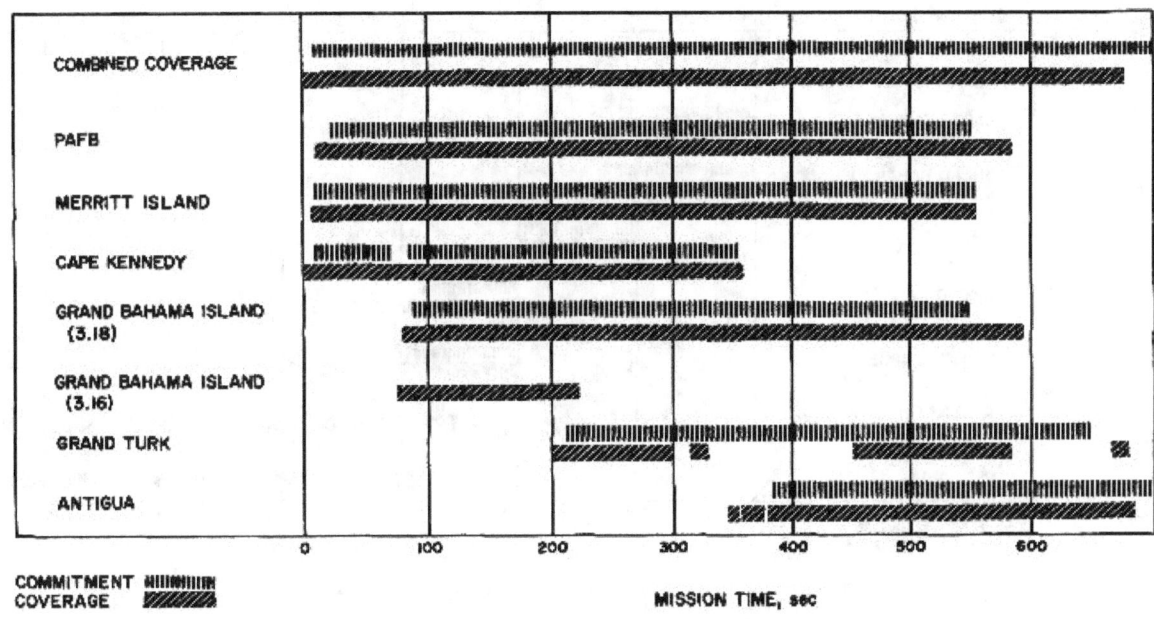

COMMITMENT COVERAGE

MISSION TIME, sec

Fig. V-2. Radar coverage, before spacecraft separation

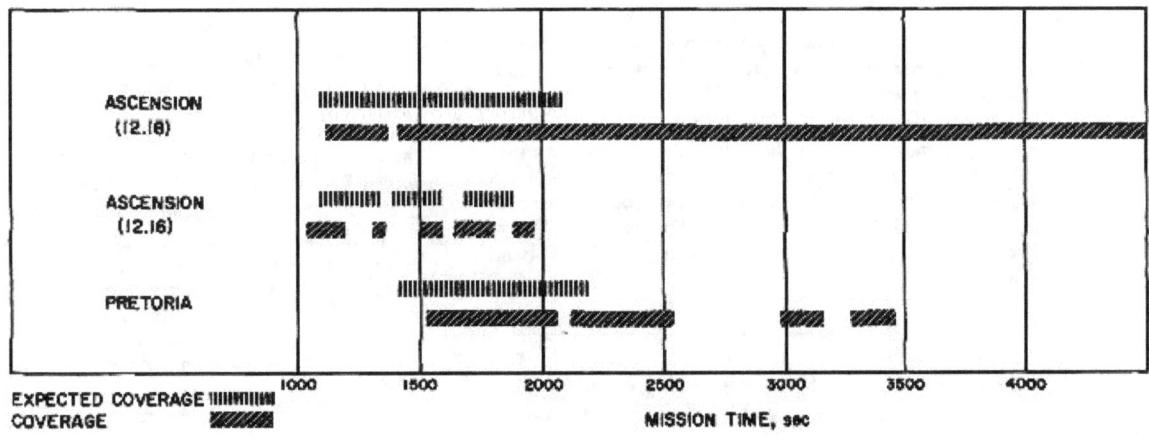

EXPECTED COVERAGE
COVERAGE

MISSION TIME, sec

Fig. V-3. Radar coverage, after spacecraft separation

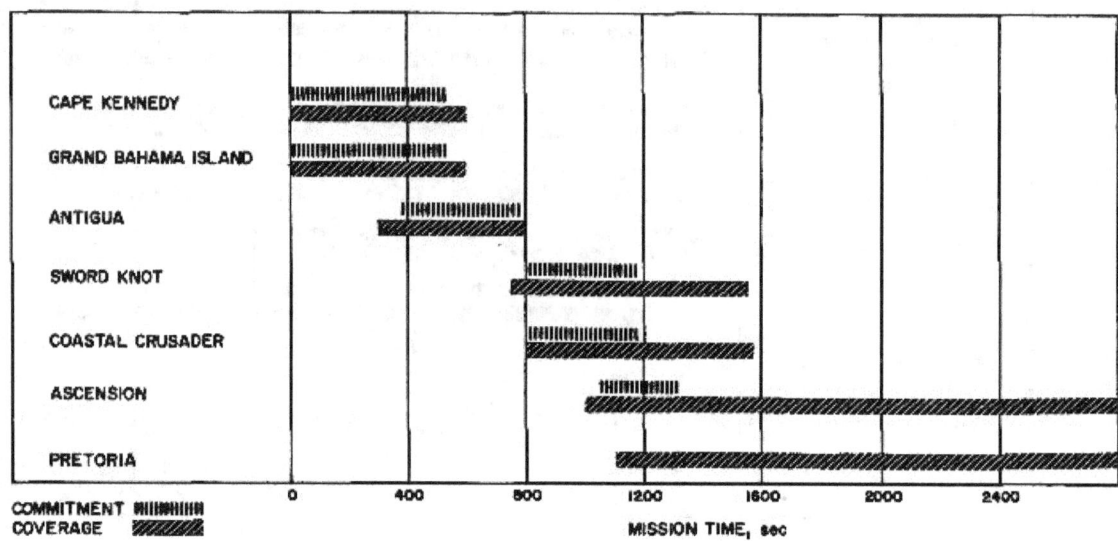

Fig. V-4. VHF telemetry coverage

The S-band telemetry resources planned to meet this requirement were the three Range Instrumentation Ships and 30-ft S-band (TAA-3A) antenna systems located at Antigua and Ascension.

All the primary S-band systems were to be used on a limited commitment basis since they were still under engineering control.

Confidence was fairly high in these systems because the *Sword Knot* had successfully provided S-band coverage for *Centaur* AC-8 launch and the *Coastal Crusader* had provided S-band support for *Pioneer*. The *General Arnold*, however, had not yet supported a live S-band mission.

A 3-ft S-band antenna, with its associated downconverter, receiver, etc., was in place at Ascension. This system provided the S-band data from Ascension for the previous *Centaur* launch and was used to back up the TAA-3A antenna system. A similar back-up system was provided at Antigua. In addition, the *General Arnold* was to cover the early interval after turn-on of *Surveyor* transmitter high power.

The TAA-3A antenna systems at Antigua and Ascension had failed to acquire data from two previous launch vehicles (the most recent was AC-8) which carried

S-band systems. RF systems engineers had been dispatched to conduct a series of tests, the results of which were analyzed to determine the cause of the problem and required action or recommendations.

The TAA-3A antennas, which occasioned so much concern prior to launch, provided excellent flight support.

The interval that was received and recorded in flight is shown in Fig. V-5. The top bar is a composite of all the stations. Data was received and recorded from $T - 4500$ to $L + 2376$ sec, with dropouts (or low-quality signals) shown that occurred over the following intervals:

$$
\begin{array}{c}
75\text{–}122 \text{ sec} \\
240\text{–}413 \\
1623\text{–}1671 \\
1741\text{–}1796 \\
2033\text{–}2290
\end{array}
$$

The *General Arnold* did not record any data of usable quality. There was a brief interval of receiver lock-up, but no time-division-multiplex lock was obtained. Postflight investigation revealed a connector with a bent center pin in the coaxial line from the antenna assembly to the preamplifiers.

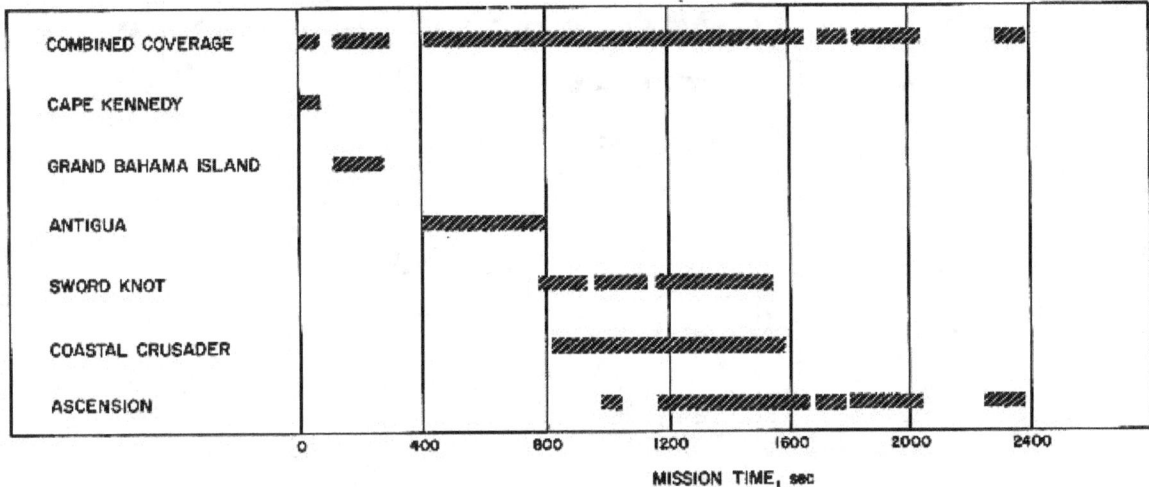

Fig. V-5. S-band telemetry coverage

4. *Surveyor* Real-Time Data

The AFETR also retransmits *Surveyor* data (VHF or S-Band) to Building AO, Cape Kennedy, for display and for retransmission to the SFOF. In addition, downrange stations monitor specific channels and report events via voice communication.

For Mission A existing hardware and software facilities were utilized to meet the real-time data requirements.

The S-band facilities participated in the Operational Readiness Tests to demonstrate capability to process and retransmit data. The *General Arnold* was the only station that did not perform satisfactorily in these tests. The *General Arnold* S-band capabilities were unknown, and the tests provided only a measure of ability to retransmit spacecraft telemetry data from a magnetic tape provided by JPL. Because of equipment problems, there were only two tests in which telemetry data retransmission from the *General Arnold* was good. Furthermore, during one of those tests, it was necessary to hold for approximately 5 hr before the *General Arnold* was able to transmit data at all.

The other AFETR stations performed satisfactorily in the tests. It should be remembered that the tests did not demonstrate station ability to acquire an S-band signal. However, with the exception of Antigua and the *General Arnold*, the other AFETR stations which were to participate in the launch had some S-band experience.

During the countdown, difficulty was encountered in establishing adequate RF communications paths for retransmission of data from Range Instrumentation Ships and Ascension.

Surveyor real-time coverage for Mission A is shown in Fig. V-6. The top bar shows the spacecraft data transmitted in real-time to Building AO. Data were transmitted until $T + 2376$ sec, with gaps shown as listed below:

Dropouts	Poor quality (> 1 in 10^3 bit error rate)
606–621 sec	1498–1514 sec
662–685	1534–1555
765–788	1619–2119
941–1000	2319–2376
2119–2319	

Neither the *General Arnold* nor the *Coastal Crusader* transmitted any data.

There was also a requirement to retransmit *Centaur* stage VHF link data to Cape Kennedy from Antigua and Grand Bahama. This requirement was not met at Antigua owing to an operator patching error, but because of the overlapping coverage given from Grand Bahama, only 84 sec of data were lost.

5. Real-Time Computer Services

The required real-time computer system services were to be met in a routine manner by the existing facilities

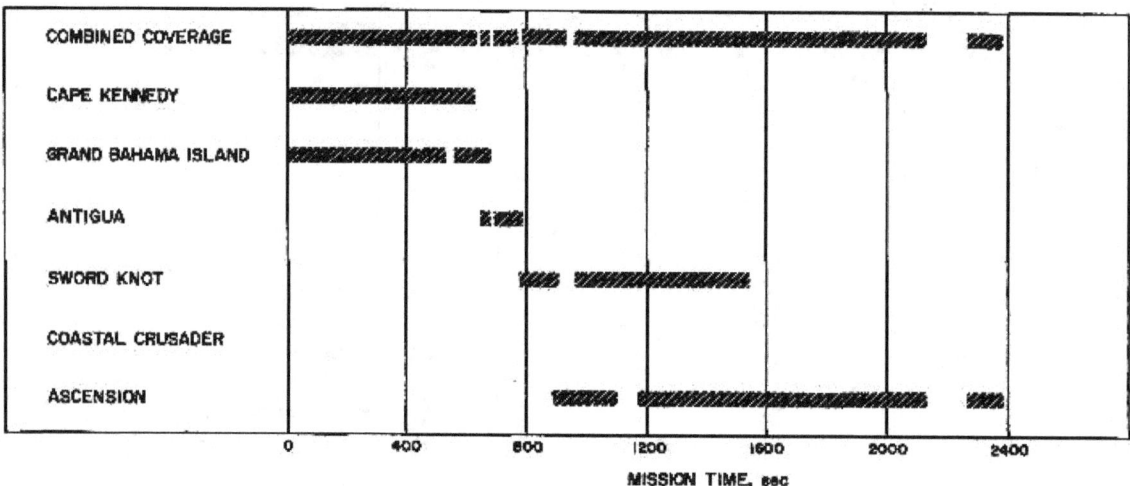

Fig. V-6. Real-time data transmission

of the real-time computer system. Although no problems occurred in this area, the computer was unable to compute a meaningful orbit until Ascension data were available. Ascension rise occurred after the start of the retromaneuver. An orbit was then computed from Ascension data, and DSS 51 predicts based on this orbit were generated.

The AFETR generated the DSS 72 predicts, and the interrange vector (IRV) message was transmitted to the DSS. Considering the limited metric data available, the real-time computer system met requirements well.

B. Goddard Space Flight Center

The Manned Space Flight Network (MSFN), managed by GSFC, supports *Surveyor* missions by performing the following functions:

1. Tracking of the *Centaur* beacon (C-band) for approximately 3.5 hr.

2. Receiving and recording telemetry of the *Centaur* links from Bermuda acquisition of signal until loss of signal at Kano.

3. Providing real-time confirmation of certain "mark" events (see Appendix A).

4. Providing real-time reformatting of Carnarvon radar data at the GSFC from the Carnarvon hexidecimal system to the 38-character octal format and retransmitting these data to the RTCF at the AFETR.

5. Providing NASCOM support to all NASA elements for simulations and launch and extending this communications support as necessary to interface with the combined worldwide network.

The GSFC supported Mission A with the equipment indicated in Table V-2.

Mission A simulation tests of the GSFC support facilities were conducted on May 16 and 23, 1966, in accordance with the Network Operations Plan. The MSFN and all of its interfacing elements were "go" at liftoff except for the Carnarvon radar, which reported parametric amplifier problems. Carnarvon, however, anticipated that this would not affect their support.

1. Acquisition Aids

Bermuda, Canary Island, and Kano are equipped with acquisition aids to track the vehicle and provide RF inputs to the telemetry receivers. Performance recorders are used to record AGC and angle errors for postmission analysis. The acquisition aids provide telemetry RF inputs from Bermuda acquisition of signal through loss of signal at Kano. All MSFN acquisition aid systems performed their required functions. There were no equipment failures and no discrepancies during the operation.

Acquisition aid coverage was good at Bermuda and Canary Island, where the vehicle was in range, and poor

Table V-2. GSFC Network configuration

Location	Acquisition	Telemetry	C-band radar	SCAMA	Telemetry	Radar high-speed data	Real-time readouts
Bermuda	X	X	X	X	X	X	X
Canary Island	X	X		X	X		X
Kano	X	X		X	X		X
Carnarvon	X		X	X	X		
GSFC				X	X	X	
Cape Kennedy				X	X	X	

at Kano and Carnarvon, where excessive range was encountered. Autotrack was not attempted at any of the MSFN stations because of low elevation angles.

Nominal pointing data generated by GSFC computers was supplied to Carnarvon only, since the data did not contain view times for Canary Island and Kano. During the minus count, GSFC computers generated and furnished pointing data to Canary Island, Kano, and Carnarvon. This pointing data was not valid, since it was generated for an 84-deg launch azimuth and the actual launch azimuth was 102.285 deg. Real-time acquisition messages were sent to Carnarvon only. These messages were furnished by RTCS computers and transmitted late in the pass.

At Carnarvon, acquisition was reported, but later indications received by the Carnarvon radar proved that this signal was in fact spurious. With the exception of the invalid pointing data, low elevation angles, and excessive range, overall performance was considered good.

2. Telemetry Data

Bermuda, Canary Island, and Kano were also equipped to decommutate, receive, and record telemetry. Capability for coverage was provided from Bermuda acquisition of signal through loss of signal at Kano. "Mark event" (see Appendix A) readouts were required from all stations in real-time or in as near real-time as possible when the vehicle was in view of a station.

MSFN telemetry support was well performed. There were no equipment failures or discrepancies during the operation. Bermuda was required to confirm Mark Events 6 through 11 on the Centaur link. The fact that Mark Events 8 through 11 were not confirmed was due

to loss of signal not station malfunctions. The only problem encountered at Bermuda was in the decommutation of commutated channel 9 on the Centaur link (RSC/AGC voltages) due to degradation by RF link noise. Except for the decommutation problem, telemetry coverage at Bermuda was considered good.

Canary Island was required to confirm Mark Events 17 through 19. Mark Event 17 was not confirmed because the event occurred prior to Canary Island acquisition of signal, while Mark Events 18 and 19 were confirmed as required. Mark Event 17 could have been lost because of the invalid pointing data received at Canary Island. Canary Island reported that decommutator lock-up was not accomplished in time to confirm Mark Event 18 in real-time due to a polarity problem in the channel wave train.

Except for the polarity problem, telemetry coverage at Canary Island was considered good.

Kano was required to confirm Mark Events 17 through 19 on the Centaur link. None of these mark events was confirmed. Because of the launch azimuth, the vehicle passed far south of Kano, at an extreme range which altered the anticipated view time considerably. Maximum signal strength was low, and therefore neither decommutation lock-on nor valid data were obtained. Although no equipment problems were encountered, telemetry coverage at Kano was considered poor.

3. Tracking Data (C-Band)

Bermuda provided radar beacon tracking, magnetic tape recording (at a minimum of 10 points/sec), and real-time data transmission to GSFC and AFETR. Carnarvon did not provide usable data owing to the problems outlined in Section 1 above.

4. Computer Support

The GSFC Data Operations Branch provided computer support during the prelaunch, launch, and orbital phases of the mission, but performance in this area was poor. Because of erroneous computer input, the nominal pointing data generated did not include view times for Canary Island and Kano. New pointing data was generated for an 84-deg launch azimuth; but the launch azimuth was 102.285 deg, and it was then too late to generate corresponding pointing data. The postlaunch data requested by LeRC was supplied late because of conflicts with *Gemini* commitments at that time.

5. Data Handling and Ground Communications

Data was provided by MSFN stations at Bermuda, Canary Island, Kano, and Carnarvon in accordance with the requirements of the Network Operations Plan. Kano failed to record the 100-kc reference on telemetry tape. Otherwise, the supporting stations recorded and shipped data within acceptable time frames and in good condition.

Existing NASCOM and DOD Network facility voice and teletype circuits provided ground communications to all participating stations.

C. Deep Space Network

The DSN supports *Surveyor* missions with the integrated facilities of the Deep Space Instrumentation Facility (DSIF), the Ground Communication System (GCS), and the DSN facilities in the Space Flight Operations Facility (SFOF).

1. The DSIF

The following Deep Space Stations, all having 85-ft antennas, were committed as prime stations for the support of Mission A.

DSS 11   Pioneer, Goldstone DSCC, Barstow, California (Fig. V-7)

DSS 42   Tidbinbilla, Australia (near Canberra)

DSS 51   Johannesburg, South Africa

These prime stations were committed to provide tracking coverage on a 24-hr/day basis, from launch to lunar landing, and for the first lunar day and night. For succeeding lunar days and nights, the commitment was for 24-hr/earth day coverage during the first three and last two earth days and for 10-hr/earth day in between.

In addition to the basic support provided by prime stations, the following facilities support was provided for Mission A:

1. DSS 71, Cape Kennedy, was committed to support RF compatibility tests with *Surveyor I* on the launch pad.

2. DSS 61, Madrid DSCC, and DSS 72 were implemented to track on an engineering basis but were not committed to provide Mission A support.

3. Facilities support was provided for mission-dependent equipment, including space for CDC's at all prime stations and for the TVGDHS at DSS 11, Goldstone DSCC. (The mission-dependent equipment is described in Section VI.)

4. Maintenance and operation was provided for all mission-independent equipment. Logistic and spares support was provided for mission-dependent equipment.

Data is handled by the prime DSIF stations as follows:

1. Tracking data, consisting of antenna pointing angles and doppler (radial velocity) data, is supplied in near-real-time via teletype to the SFOF and post-flight in the form of punched paper tape. Two-way and three-way doppler data is supplied full-time during the lunar flight, and also during lunar operations at *Surveyor* Project Office request. The two-way doppler function implies a transmit capability at the prime stations.

2. Spacecraft telemetry data is received and recorded on magnetic tape. Baseband telemetry data is supplied to the Command and Data Handling Console (CDC) for decommutation and real-time readout. The DSIF also performs precommunication processing of the decommutated data, using an on-site data processing (OSDP) computer. The data is then transmitted to the SFOF in near-real-time, using high-speed data modems.

3. Video data is received and recorded on magnetic tape. This data is sent to the CDC, and, at DSS 11, Goldstone DSCC, only, to the TV Ground Data Handling System (TVGDHS; TV-11) for photographic recording. In addition, video data from DSS 11 is sent in real-time to the SFOF for magnetic and photographic recording by the TVGDHS (TV-1).

Fig. V-7. DSS 11, Goldstone DSCC, Master Station of the DSIF

After lunar landing, DSS 11 performs a special function. Two receivers are used for different functions. One provides a signal to the CDC, the other to the TVGDHS. (Signals for the latter system are the prime *Surveyor* Project requirement during this phase of a mission.)

4. Command transmission is another function provided by the DSIF. Approximately 280 commands to the

spacecraft are made during the nominal sequence from launch to touchdown. Confirmation of the commands sent is processed by the OSDP computer and transmitted by teletype to the SFOF.

The characteristics for the S-band and L/S-band antenna systems are given in Table C-3.

Table V-3. Characteristics for S-Band and L/S-Band antenna systems

Station configuration	GSDS S-band	L/S conversion	Station configuration	GSDS S-band	L/S conversion
Antenna			**Transmitter characteristics**		
Tracking	85-ft parabolic	85-ft parabolic	Frequency		
Mount	Polar (HA-Dec)	Polar (HA-Dec)	(nominal)	2113 mc	2113 mc
Beamwidth ±3 db	~0.4 deg	~0.4 deg	Frequency channel	14b	14b
Gain, receiving	53.0 db, +1.0, −0.5	53.0 db, +1.0, −0.5	Power, max.	10 kw	10 kw
Gain, transmitting	51.0 db, +1.0, −0.5	51.0 db, +1.0, −0.5	Tuning range	±100 kc	±100 kc
Feed	Cassegrain	Cassegrain	**Modulator, phase**		
Polarization	LH** or RH circular	RH circular	Input impedance	$\geq 50\ \Omega$	$\geq 50\ \Omega$
Max. Angle tracking rate*	51 deg/min = 0.85 deg/sec	51 deg/min = 0.85 deg/sec	Input voltage	≤ 2.5 v peak	≤ 2.5 v peak
Max. angular acceleration	5.0 deg/sec/sec	5.0 deg/sec/sec	Frequency response (3 db)	DC to 100 kc	DC to 100 kc
Tracking accuracy (1σ)	0.14 deg	0.14 deg	Sensitivity at carrier	1.0 rad peak/v peak	1.0 rad peak/v peak
			Output frequency		
Antenna Acquisition	2 × 2-ft horn	2 × 2-ft horn	Peak deviation	2.5 rad peak	2.5 rad peak
Gain, receiving	21.0 db ±1.0	21.0 db ±1.0	Modulation deviation	±5%	±5%
Gain, transmitting	20.0 db ±2.0	20.0 db ±2.0	**Stability**		
Beamwidth ±3 db	~16 deg	~16 deg	Rubidium standard	Yes	Yes
Polarization	RH circular	RH circular	Stability, short-term (1σ)	1×10^{-11}	1×10^{-11}
Receiver	S-band	S-band	Stability, long-term (1σ)	5×10^{-11}	5×10^{-11}
Typical system temperature	—	—	Doppler accuracy at F_{rc} (1σ)	0.2 cps = 0.03 m/sec	0.2 cps = 0.03 m/sec
With paramp	270 ±50°K	270 ±50°K	**Data transmission, teletype**		
With maser	55 ±10°K	55 ±10°K	Angles	Near-real-time	Near-real-time
Loop noise bandwidth	—	—	Doppler	Near-real-time	Near-real-time
Threshold (2B$_{LO}$)	12, 48 or 152 cps +0, −10%	12, 48 or 152 cps +0, −10%	Telemetry	N/A	N/A
Strong signal (2B$_{LO}$)	120, 255, or 550 cps +0, −10%	120, 255, or 550 cps +0, −10%	**Command and data handling Console (CDC)**		
			Channel two		
Frequency (nominal)	2295 mc	2295 mc	Discriminator	Yes	Yes
			Command	N/A	N/A
Frequency channel	14 amp	14 amp	capability		

*Both axes.
**Goldstone only.
DSS 72 30-ft Az-El Ant characteristics differ from 85-ft HA-Dec Ants.

The angle tracking parameters for all stations are as follows:

1. Maximum angle tracking rate (both axes)........ 51 deg/min = 0.85 deg/sec

2. Maximum angular acceleration .... 5.0 deg/sec/sec

3. Tracking accuracy (one standard deviation) .... $\sigma = 0.14$ deg

4. The system doppler tracking accuracy at the receiver carrier frequency for one standard deviation .... 0.2 cps = 0.03 m/sec
The maximum doppler tracking rate depends on the loop noise bandwidth, and for phase error of 30 deg and strong signal (-100 dbm) it is as follows:

Loop noise bandwidth, cps	Maximum tracking rate, cps/sec
12	100
48	920
152	5000

The receiver characteristics for S-band and L/S-band stations are as follows:

1. *Noise temperature.* The total effective system noise temperature including circuit losses when looking at or near the galactic pole is:
   Traveling-wave masers, $55 \pm 10°K$
   Parametric amplifier, $270 \pm 50°K$

2. *Loop noise bandwidth.* The closed-loop noise bandwidth for various signal conditions is:
   Threshold $2B_{L0}$, 12, 48, or 152 cps +0%
   $\qquad\qquad\qquad\qquad\qquad\qquad -20\%$
   Strong signals $2B_1$, 132, 274 or 518 cps +0%
   $\qquad\qquad\qquad\qquad\qquad\qquad -20\%$

3. *Threshold.* Carrier lock will be maintained with an rms phase error due to noise of less than 30 deg when the ratio of carrier power to noise power in the closed-loop noise bandwidth is 6 db or greater. Owing to the nature of the operation of the phase-lock loop, this condition requires the carrier power at the receiver input to be 9 db greater than the value at threshold, which is defined as a carrier-to-noise power ratio of zero db in the threshold loop noise bandwidth $2B_{L0}$.

a. DSIF preparation testing. The *Surveyor* Project Mission A Test Plan for the three prime stations is shown in Table V-4. Three series of tests (A, B, and C) are included in this plan. These tests as well as other tests required for Mission A are described below:

1. *A-Tests.* These are facility preparation and training tests which are scheduled and conducted individually by each station (there is no GCS or SFOF participation). All prime stations completed a full sequence of A-Tests.

Table V-4. Mission A Test Plan summary for DSS 11, 42, and 51

Test		Hours required	Week test conducted (1966)
DSS 11			
Engineering Verification		10	1-31
Operations Verification		20	1-31
SFOF/DSIF Compatibility and	B-1.1	35	2-7
TV System	B-3.0		
Lunar Sequence Testing	C-1.1	20	2-14
			3-28
SFOF/DSIF Compatibility and,	B-1.1	35	4-4
TV System	B-3.0		
SFOF/DSIF Operational Test	C-1.1	19	4-25
Engineering Verification		40	5-2
SFOS Integration	C-3.0	20	5-9
SFOS Integration	C-3.0	20	5-16
SFOS Operational Readiness	C-5.0	20	5-23
DSS 42			
Engineering Verification		10	1-24
Operations Verification		20	1-24
SFOF/DSIF Compatibility	B-1.3	26	1-31
SFOF/DSIF Operational Test	C-1.3	16	5-2
Engineering Verification		40	4-25
SFOS Integration	C-3.0	12	5-9
SFOS Integration	C-3.0	12	5-16
SFOS Operational Readiness	C-5.0	12	5-23
DSS 51			
Engineering Verification		10	1-17
Operations Verification		20	1-17
SFOF/DSIF Compatibility	B-1.2	25	1-24
SFOF/DSIF Operational Test	C-1.2	18	4-11
Engineering Verification		40	5-2
SFOF/DSIF Operational Test	C-1.2	18	4-18
SFOS Integration	C-3.0	13	5-9
SFOS Integration	C-3.0	13	5-16
SFOS Operational Readiness	C-5.0	13	5-23

2. *B-Tests.* These are functional compatibility tests designed to ensure that ground-based facilities are capable of processing telemetry data (and video data at DSS 11, Goldstone DSCC, only) as received from the spacecraft. Command capability is also verified in all configurations and modes of operation. During these tests, data is sent from the prime stations to the SFOF for processing, and full mission support in the SFOF is usually required.

Many last-minute change requests were received from the Project Office after completion of the B-Tests. These changes were accommodated to the maximum extent possible, and all changes which the Project Office designated as "Mission critical" were accomplished. On launch day, the DSN was ready to support the *Surveyor* Mission A. But in the period between the preflight review (May 4, 1966) and launch there were some unforeseen developments. The most significant of these, in terms of effect on pre-mission preparation, was an electrical fire in the basement of the SFOF on Friday, May 13, 1966. Recovery actions included rewiring of the power circuits (Fig. V-8) and cleanup and checkout of the computers to ensure that they were fully operational. The net effect on pre-mission preparation was that two integration tests were cancelled. (One of these

tests was rescheduled and performed six days after it was originally scheduled.)

3. *C-Tests.* These are operational tests to verify that all prime stations, communications, and the SFOF are fully prepared to meet Mission responsibility. Selected portions of the Sequence of Events are followed rigidly, using both standard and nonstandard procedures.

DSS 11, 42, and 51 participated in the Operational Readiness Test (ORT), C-5.0, which was conducted during the week prior to launch. In addition, DSS 61 and 72 participated in the ORT for engineering practice. An evaluation of station and net control support during the ORT indicated the readiness of the T&DA System.

4. *Training Station Tests.* The *Surveyor* Project had no pending requirements for any of the DSIF training stations, provided Mission A was launched during the May/June window. As a result, no B- or C-Tests were scheduled for these stations by the *Surveyor* Project. However, A-Tests were scheduled for DSS 61 (during the week of April 25) and for DSS 72 (during the week of May 2).

5. *T-21 Compatibility and Training Tests.* Compatibility tests were run between the test spacecraft

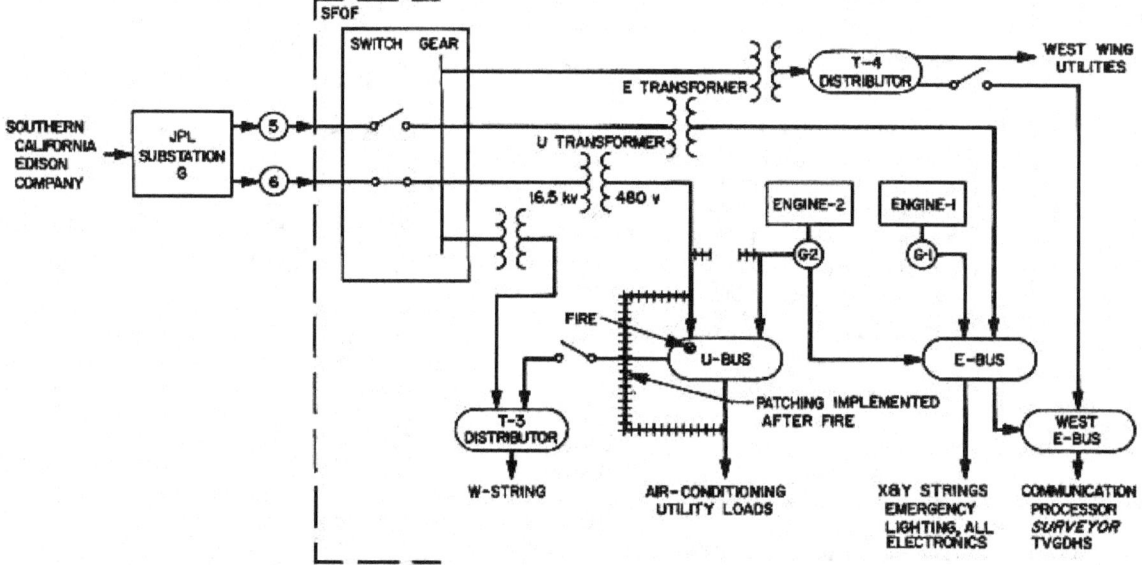

Fig. V-8. SFOF power system rewiring

(T-21) and DSS 71 in July 1965 at AFETR. Compatibility was demonstrated in a gross way, and an anomaly found between the STEA and DSS frequency measurements was corrected.

A series of tests between the T-21 spacecraft and DSS 11, Goldstone DSCC, was conducted beginning in August 1965. These tests involved the T-21, DSS 11, the CDC, and the SFOF, and established mutual compatibility between all of these elements of the network. The tests included RF tests, command tests, and telemetry tests and were completed in October 1965.

Another series of T-21 tests began at Goldstone DSCC in March 1966 for training of operator crews for DSS 61 and DSS 72. T-21 was also used for SSAC/SPAC lunar sequence training and as a data source for B- and C-Tests.

6. *Surveyor On-Site Computer Program Integration Tests.* These tests were designed to check out the *Surveyor* on-site computer program (SOCP) and to verify that data can be transmitted from a DSIF station to the SFOF and be processed there. Such tests were run on a regular basis with each prime station (DSS 11, 42, and 51). These tests were concluded with a checkout of the final Mission A version of the SCOP program in April 1966.

b. DSIF flight support. All of the DSIF prime and engineering practice stations reported "go" status during the countdown. All measured station parameters were within nominal performance specifications and communications circuits were up.

Figure V-9 is a profile of the DSIF mission activity up to lunar touchdown. This figure contains the periods

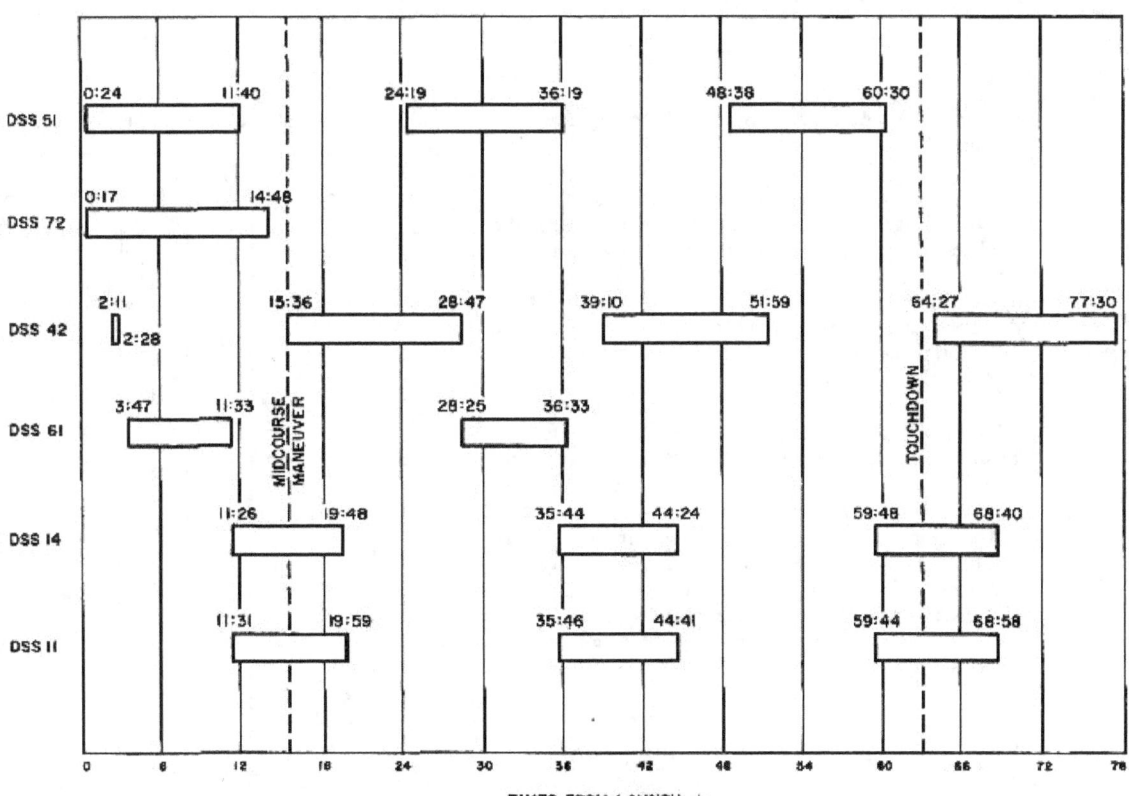

TIMES FROM LAUNCH, hr

Fig. V-9. Station view periods

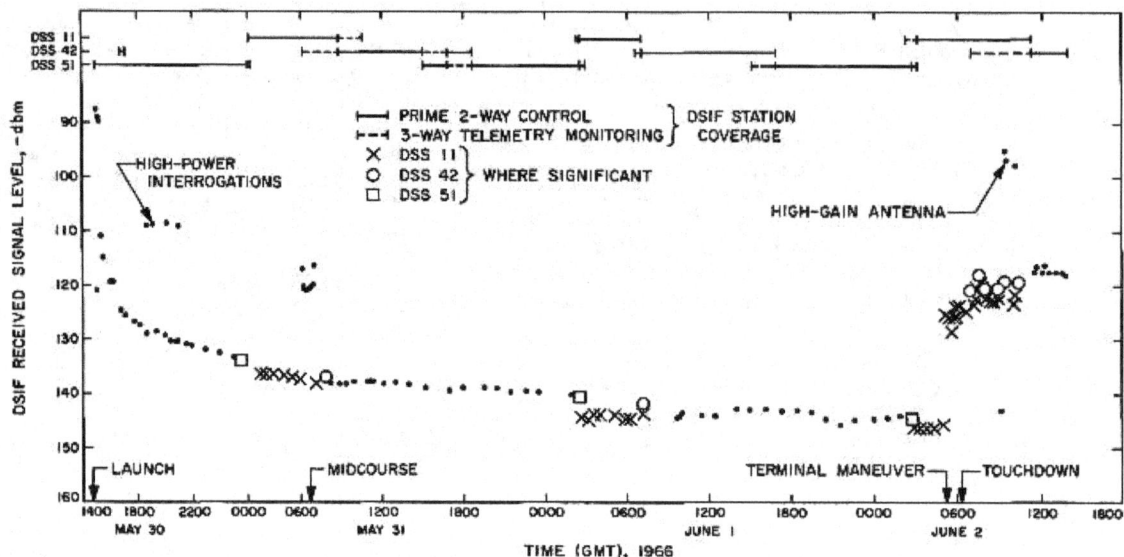

Fig. V-10. DSS-received signal level

each station tracked the spacecraft plotted against time in hours.

The signal levels received at the prime stations are shown in Fig. V-10. They compare favorably with the predicted values. A 2-db discrepancy between the reported levels from DSS 11, Goldstone DSCC, and the other stations is apparent. This data is compiled from the tracking and post-tracking reports and is therefore not continuous, but it is probably sufficient to indicate performance in general.

Table V-5 shows the DSIF tracking passes for the Mission. Table V-6 is a tabulation of commands transmitted by the prime stations during each spacecraft pass.

Throughout the mission, from launch through touchdown to the end of the first lunar day, all stations remained within performance specifications except where significant anomalies caused the complete shutdown of the equipment affected. The anomalies encountered during Mission A are shown in Table V-7.

During the critical touchdown sequence, both DSS 11, Goldstone DSCC, receivers remained in lock, using loop bandwidths of 152 and 12 cps, respectively. The role of the DSN after touchdown is well portrayed by its support of the TV effort, as shown in Table V-8.

Table V-5. DSIF tracking passes

Pass	DSS station						
	11	14	42	51	61	71	72
0			X	[X]		X	[X]
1	X	X	X	[X]	X		[X]
2	X	X	X	X	X		
3	X	X	X	X			
4	X		X	X			
5	X		X	X			
6	X		X	X			
7	X		X	X			
8	X		X	X			
9	X		X	X			
10	X		X	X			
11	X		X	X			X
12	X		X	X	X		
13	X		X	X			X
14	X		X	X	X		
15	X		X	X			
16	X		X	X			
17	X			X			
Totals	17	3	17	17	4	1	3

[ ] Indicates one continuous pass.

Table V-6. Commands transmitted by prime tracking stations
(not including second lunar day operations)

Pass No.	Date (1966 GMT)	Event	Commands transmitted		
			DSS 51	DSS 11	DSS 42
1	May 30/31	Midcourse commands sent by DSS 11	77	84	3
2	May 31/ June 1		10	8	26
3	June 1/2	Soft landing, 06:17:37 GMT (June 2)	11	56	590
4	June 3		30	4833	6
5	June 3/4		60	1080	680
6	June 4/5		5	4372	988
7	June 5/6		20	7876	300
8	June 6/7	Lunar noon, 06:17 GMT (June 7)	1656	9345	67
9	June 7/8		30	3158	18
10	June 8/9		20	4000	15
11	June 9/10		15	12,000	17
12	June 10/11		22	10,100	71
13	June 11/12		237	10,611	35
14	June 13		64	7614	444
15	June 14/15	Start lunar night, 15:12 GMT (June 14)	25	5000	109
16	June 15/16		73	96	30
17	June 16	Spacecraft power turned off by DSS 11, 16/20:31:00 GMT	80	100	N/A
		Station Total	2435	80,333	3399
		Grand Total		86,177	

Note: Of 86,177 commands transmitted, only one (sequence 0316) was rejected by the spacecraft.

Table V-7. Summary of DSIF anomalies

DSS	Pass	Time	Anomaly	Probable cause	Immediate reaction	Effect on mission	Current action	Comments
51	1	1521	Rejection of Command 0316, Major Sequence 0040	Insertion of monitor jack on FR 1400 recorder	Retransmit command	Not affected	Stations to be advised of correct operational procedure. Additional recorder modules requested to eliminate need for patching.	Retransmitted command accepted. Circuit interaction being investigated. Availability of nonshorting jacks being investigated
51	1	1545	Receiver drops lock	VCO instability caused by circuit breaker trip in receiver rack	Spare VCO installed	Intermittent data for 8 min	Investigate problem under TFR 51-RCV-029	Suspect overload due to additional monitor equipment
51	1	1725	Dec angle hitch on antenna	Insufficient servo bandwidth	Increase Dec bandwidth to .05 cps	Temporary reduction in signal level	Specifically correct servo bandwidths in the future	Station has to use discretion to cope with real-time conditions
11	1	0312	Noisy doppler data	Partial failure of doppler counter	Replace module B.15 in Counter No. 2	70 min of lost tracking data		Random failure
42	1	1645	No telemetry data on HS data line	Faulty interrupt from HSDL modem or defect in SOCP	Restart program and change from 910 to 920	Temporary loss of HS data	Investigate problem under TFR 42-DIS-033-5	Random failure
51	2	1758	Failure of heat exchanger temperature regulator	Tube failure	Replace tube	Transmitter off for 58 min	Investigation proceeding under TFR-51-TXR-021	DSS 42 overlap
11	2	0226	Unable to maintain good two-way doppler data	Defective coax relay in exciter VCO selector	Bypass coaxial relay	Loss of tracking data	Document fix	Early transfer to DSS 42
42	4		Defective maser	Leaks in CCR system	Transfer to paramp for one pass	Not affected. Good video data on paramp	Repair CCR and cool down	Common failure
51	8	0128	Loss of transmitter drive	Faulty 22-mc distribution amplifier	Replace distribution amplifier	Transmitter off for 14 min	Nil	Random failure
42	11	0148	Antenna stopped	Blown 15A fuse in HA power panel	Replace fuse	Antenna stopped for 12 min	Investigate under TFR-42-ANT-011-5	Random failure
51	12	0503	Transmitter outage	Leak in safety relief valve	Top up cooling water supply and bypass relief valve	Transmitter out for 48 min	Investigate under TFR-51-TXR-024	Common failure
42	13	1965	Transmitter interlock tripped off	Leak in heat exchanger pump gasket	Repair gasket and refill tank	Delayed transfer to DSS 11	Investigate under TFR-42-TXR-012-5	Common failure
51	14		Transmitter interlock tripped off	Leak in heat exchanger pump gasket	Top up tank every 15 min to end of pass	Not affected	Investigate under TFR-51-TXR-025	Common failure
42	16		Inoperative maser	Oil leak in compressor	Transfer to paramp	Not affected	Undertake thorough maintenance of compressor	Common failure

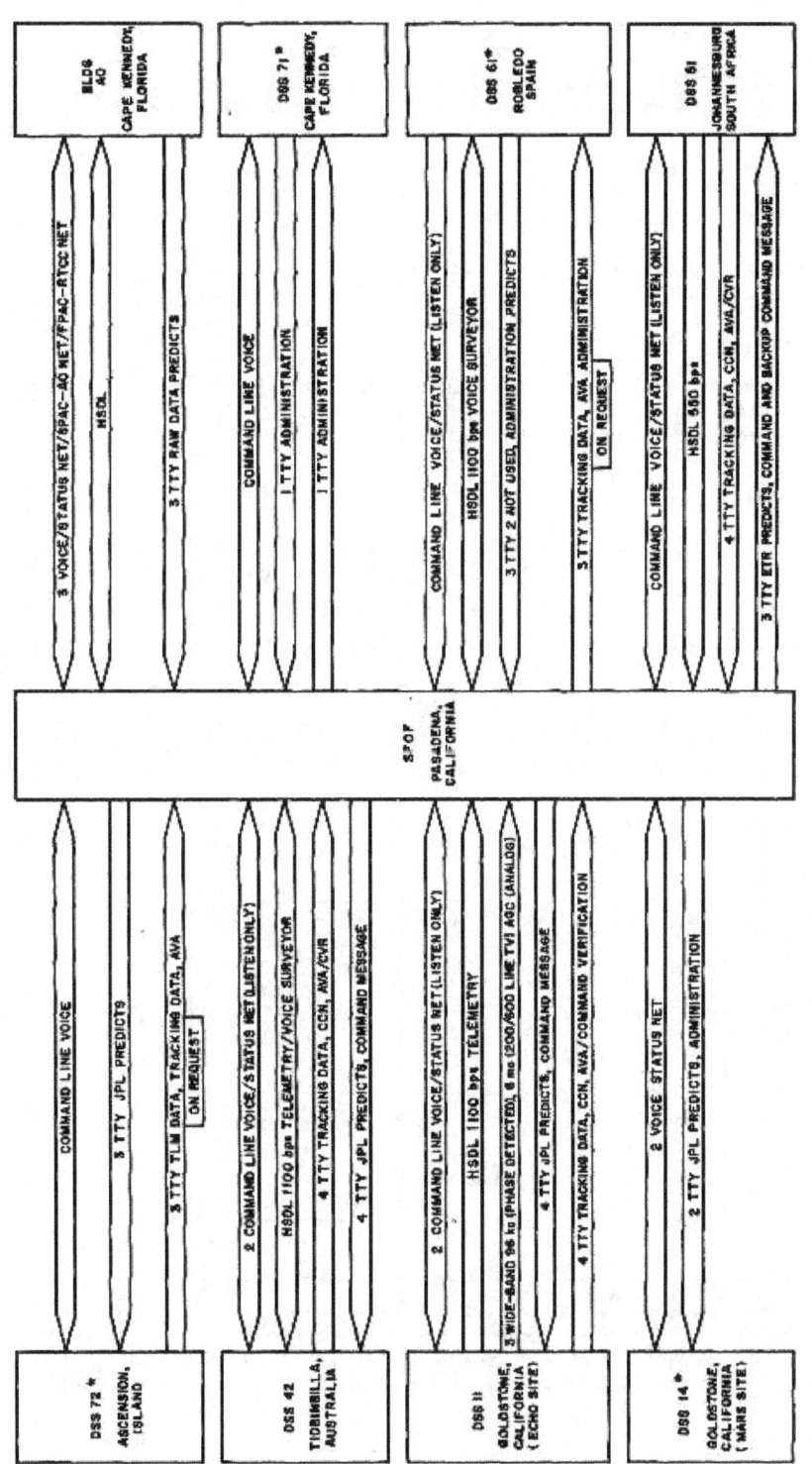

Fig. V-11. Ground communication system links

The DSIF provided the required tracking and telemetry coverage 24 hr/day, from $L + 21$ min, acquisition, through the flight and lunar operations, until spacecraft shutdown on June 16, 1966. Some major hardware problems occurred, such as doppler and transmitter problems at DSS 11, Goldstone DSCC, but the system had enough redundancy so that the overall performance of the DSIF was not seriously degraded.

DSIF operator training and documentation were excellent.

2. GCS/NASCOM

For *Surveyor* missions, the GCS transmits tracking, telemetry, and command data from the DSIF to the SFOF and control and command functions from the SFOF to the DSIF by means of NASCOM facilities. It also transmits simulated tracking data to the DSIF and video data and base-band telemetry from DSS 11, Goldstone DSCC, to the SFOF. The general arrangement of the links involved in the system is shown in Fig. V-11.

The types and quantities of lines used are:

1. Teletype (four lines available): to all prime sites for tracking data, telemetry, commands, and administrative traffic.

2. Voice (one line available): to be shared between the DSIF and *Surveyor* Project for administrative and control functions.

3. High speed data (one line available): from each site to the SFOF for telemetry data transmission in real-time.

Table V-8. TV pictures made by commands from prime tracking stations (not including second lunar day operations)

Pass No.	Date (1966 GMT)	Event	Number of TV pictures		
			DSS 51	DSS 11	DSS 42
3	June 2	Soft landing 06:17:37 GMT (June 2)		145 13 (incl 13 200-line)	
4	June 3			863	
5	June 3/4		3	395	
6	June 4/5			646	(incl 13
7	June 5/6			867	263*
8	June 6/7	Lunar noon, 06:17 GMT (June 7)		1048	
9	June 7/8				
10	June 8/9				
11	June 9/10			1759	
12	June 10/11			1373	
13	June 11/12			1363	
14	June 13			1321	
15	June 14/15	Start lunar night, 15:12 GMT (June 14)		520**	
16	June 15/16				
17	June 16	Spacecraft power turned off by DSS 11 20:31:00 GMT (June 16)			
		Station Total →	3	10,300	13
		Grand Total →		10,316	

*Received during DSS 11 transmission and DSS 11/42 mutual view period.
**Last picture was of Pad 2 by earth light (after lunar sunset). Exposure was f4.0 at 4 min.

4. Wideband microwave to DSS 11, Goldstone DSCC: two 6-Mc lines for video to the SFOF, and one duplex 96-kc line.

The GCS provided all required support for Mission A. New communications procedures requested by *Surveyor* Project Office were successfully implemented.

3. DSN in SFOF

The DSN provides a data processing system (DPS) in the SFOF which performs the following functions for *Surveyor* missions:

1. Computation of acquisition predictions for DSIF stations (antenna pointing angles and receiver and transmitter frequencies).

2. Orbit determinations.

3. Midcourse maneuver computations and analysis.

4. On-line telemetry processing.

5. Command tape generation.

6. Simulated data generation (telemetry and tracking data).

The general configuration of the DPS for the *Surveyor* A Mission is shown in Fig. V-12.

Mission control facilities are also provided in the SFOF as follows:

1. Space in mission-dependent and mission-independent areas. Mission Control Room No. 1 and Mission Support Area No. 1, which includes the Spacecraft Performance Area and the Space Science Area, are devoted exclusively to Mission A support. Flight Path Analysis Area No. 1 is a shared area which Mission A has exclusive use of during the flight. Common user areas supporting Mission A include DSIF Net Control, the Data Processing Area, and the Communications Center.

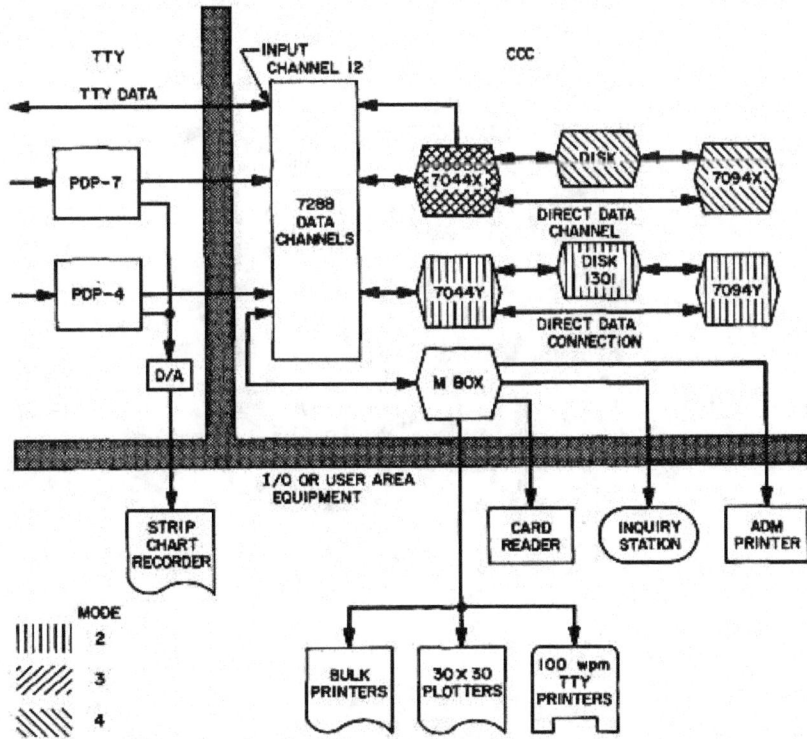

Fig. V-12. General configuration of SFOF data processing system

2. SFOF Internal Communications, including the Operational Voice Communications System (voice nets), closed circuit television, and internal teletype distribution.

3. Support, including operation of display devices, data distribution in the facility, supervision of access control, and building housekeeping functions.

4. Mission-dependent equipment space and facility support is provided for the TVGDHS (located in the west wing of the second floor of the SFOF).

The Data Processing System performed in a nominal manner with only minor hardware problems which did not detract from Mission A support.

4. DSN/AFETR Interface

This interface is supplied by the DSN to provide real-time transmission of down-range spacecraft telemetry data from Building AO at AFETR to the SFOF. The DSN is also responsible for the Range meeting the requirements for S-band telemetry coverage.

The DSN provides an interface for both VHF and S-band downrange telemetry. The nominal switchover time is after the spacecraft S-band transmitter is turned to high-power. The interface with the *Surveyor* Project is at the input to the CDC in Hangar AO. The output of the CDC is then interfaced with the Ground Communications System for transmission to the SFOF. It is also possible to go directly from the Range data output to the GCS, bypassing the CDC.

This real-time telemetry transmission interface performed quite well. Good data was received at the SFOF at all times that good data was received in Hangar AO at AFETR.

VI. MISSION OPERATIONS SYSTEM

A. Functions and Organization

The basic functions of the Mission Operations System (MOS) are the following:

1. Continual assessment and evaluation of mission status and performance, utilizing the tracking and telemetry data received and processed.

2. Determinaion and implementation of appropriate command sequences required to maintain spacecraft control and to carry out desired spacecraft operations during transit and on the lunar surface.

The *Surveyor* command system philosophy introduces a major change in the concept of unmanned spacecraft control: virtually all in-flight and lunar operations of the spacecraft must be initiated from earth. In previous space missions, spacecraft were directed by a minimum of earth-based commands. Most in-flight functions of those spacecraft were automatically controlled by an on-board sequencer which stored preprogrammed instructions. These instructions were initiated by either an on-board timer or by single direct commands from earth. For example, during the *Ranger VIII* 67-hr mission, only 11 commands were sent to the spacecraft; whereas for a standard *Surveyor* mission, approximately 280 commands must be sent to the spacecraft during the transit phase, out of a command vocabulary of 256 different direct

commands. For *Surveyor* Mission A, 288 commands were sent during transit and over 100,000 commands were sent following touchdown.

Throughout the space-flight operations of each *Surveyor* mission, the command link between earth and spacecraft is in continuous use, transmitting either fill-in or real commands every 0.5 sec. The *Surveyor* commands are controlled from the SFOF and are transmitted to the spacecraft by a DSIF station.

The equipment utilized to perform MOS functions falls into two categories: mission-independent and mission-dependent equipment. The former is composed chiefly of the *Surveyor* T&DA system equipment and has been described in Section V. It is referred to as mission-independent because it is general-purpose equipment which can be utilized by more than one NASA project when used with the appropriate project computer programs. Selected parts of this equipment have been assigned to perform the functions necessary to the *Surveyor* Project. The mission-dependent equipment (described in Section VI-B, following) consists of special equipment which has been installed at DSN facilities for specific functions peculiar to the project.

The *Surveyor* Project Manager, in his capacity as Mission Director, is in full charge of all mission operations. The Mission Director is aided by the Assistant

Mission Director and a staff of mission advisors. During the mission, the MOS organization is as shown in Fig. VI-1.

Mission operations are under the immediate, primary control of the Space Flight Operations Director (SFOD) and supporting *Surveyor* personnel. Other members of the team are the T&DA personnel who perform services for the *Surveyor* Project.

During space-flight and lunar surface operations, all commands are issued by the SFOD or his specifically delegated authority. Three groups of specialists provide technical support to the SFOD. These groups are special-

ists in the flight path, spacecraft performance, and scientific experiments, respectively.

1. Flight Path Analysis and Command Group

The Flight Path Analysis and Command (FPAC) group handles those space-flight functions that relate to the location of the spacecraft. The FPAC Director maintains control of the activities of the group and makes specific recommendations for maneuvers to the SFOD in accordance with the flight plan. In making these recommendations, the FPAC Director relies on five subgroups of specialists within the FPAC Group.

1. The Trajectory Prediction Group (TPG) determines the nominal conditions of spacecraft injection and

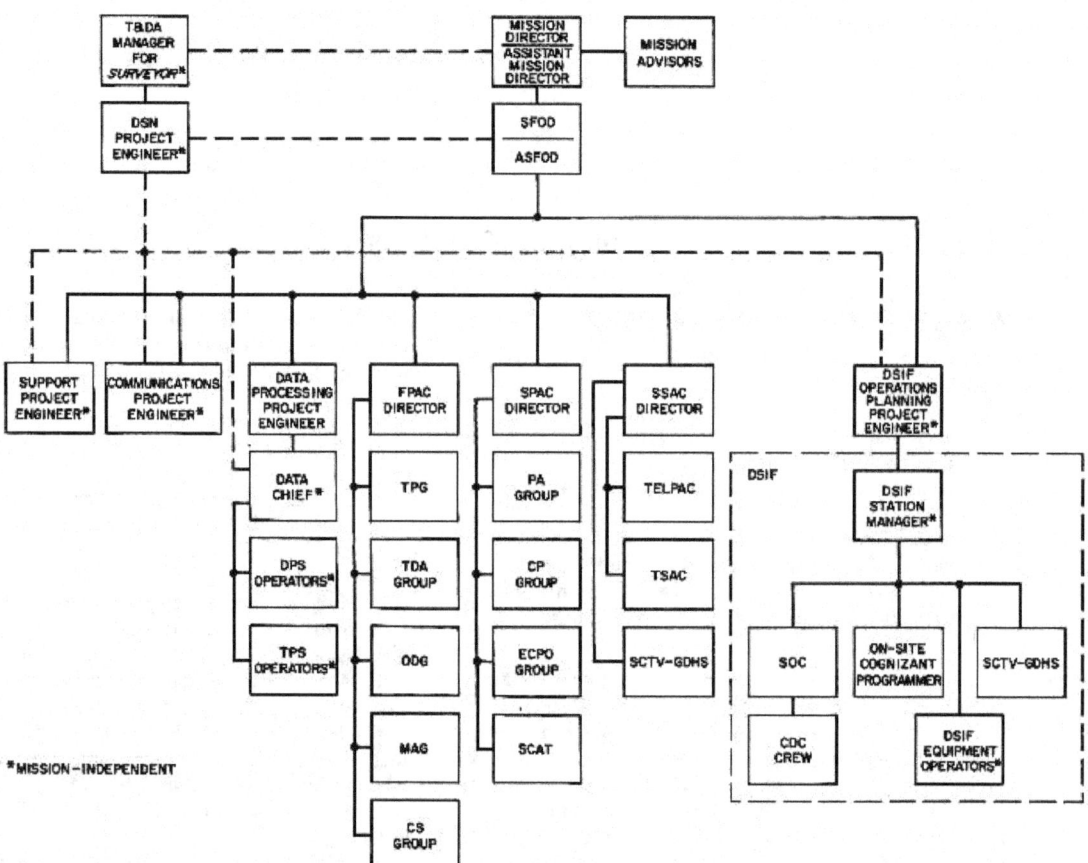

Fig. VI-1. Organization of MOS

generates lunar encounter conditions based on injection conditions as reported by AFETR and computed from tracking data by the Orbit Determination Group. The actual trajectory determinations are made by computer.

2. The Tracking Data Analysis (TDA) group makes a quantitative and descriptive evaluation of tracking data received from the DSIF stations. The TDA group provides 24-hr/day monitoring of incoming tracking data. To perform these functions the TDA group takes advantage of the Data Processing System (DPS) and of computer programs generated for their use. The TDA acts as direct liaison between the data users (the orbit determination group) and the DSIF and provides predicts to the DSIF.

3. The Orbit Determination Group (ODG), during mission operations, determines the actual orbit of the spacecraft by processing the tracking data received from the DSN tracking stations by way of the TDA group. Also, statistics on various parameters are generated so that maneuver situations can be evaluated. The ODG generates tracking predictions for the DSIF stations and recomputes the orbit of the spacecraft after maneuvers to determine the success of the maneuver.

4. The Maneuver Analysis Group (MAG) is the subgroup of FPAC responsible for describing possible midcourse and terminal maneuvers for both standard and nonstandard missions in real-time during the actual flight. In addition, once the decision has been made as to what maneuver should be performed, the MAG generates the proper spacecraft commands to effect these maneuvers. These commands are then relayed to the Spacecraft Performance Analysis and Command Group to be included with other spacecraft commands. Once the command message has been generated, the MAG must verify that the calculated commands are correct.

5. The Computing Support Group acts in a service capacity to the other FPAC subgroups, and is responsible for ensuring that all computer programs used in space operations are fully checked out before mission operations begin and that optimum use is made of the Data Processing System facilities.

2. Spacecraft Performance Analysis and Command Group

The Spacecraft Performance Analysis and Command (SPAC) Group, operating under the SPAC Director, is basically responsible for the operation of the spacecraft itself. The SPAC Group is divided into three subgroups:

1. The Performance Analysis (PA) group monitors incoming engineering data telemetered from the spacecraft, determines the status of the spacecraft, and maintains spacecraft status displays throughout the mission. The PA group also determines the results of all commands sent to the spacecraft. In the event of a failure aboard the spacecraft, as indicated by telemetry data, the PA group analyzes the cause and recommends appropriate nonstandard procedures.

2. The Command Preparation (CP) group is basically responsible for preparing command sequences to be sent to the spacecraft. In so doing they provide inputs for computer programs used in generating the sequences, verify that the commands for the spacecraft have been correctly received at the DSS, and then ascertain that the commands have been correctly transmitted to the spacecraft. If nonstandard operations become necessary, the CP group also generates the required command sequences.

3. The Engineering Computer Program Operations (ECPO) group includes the operators for the DPS input-output (I/O) console and related card punch, card reader, page printers, and plotters in the spacecraft performance analysis area (SPAA). The ECPO group handles all computing functions for the rest of the SPAC group, including the maintenance of an up-to-date list of parameters for each program.

In order to take maximum advantage during the mission of the knowledge and experience of the various personnel who are not a part of the "hard-core" operations teams (FPAC, SPAC, and SSAC) but have been engaged in detail design, analysis, or testing of the spacecraft, a Spacecraft Analysis Team (SCAT) has been established. The SCAT group, located in a building adjacent to the SFOF, has appropriate data displays showing the current status of the mission. The SCAT is available upon request for immediate consultation and detailed analysis in support of the SPAC.

3. Space Science Analysis and Command Group

The Space Science Analysis and Command (SSAC) group performs those space-flight functions related to the operation of the survey TV camera. SSAC is divided into two operating sub-groups:

1. The Television Performance Analysis and Command (TelPAC) group analyzes the performance of the

TV equipment and is responsible for generating standard and nonstandard command sequences for the survey TV cameras.

2. The Television Science Analysis and Command (TSAC) group analyzes and interprets the TV pictures for the purpose of ensuring that the mission objectives are being met. The TSAC group is under the direction of the Project Scientist and performs the scientific analysis and evaluation of the TV pictures.

The portion of the spacecraft TV Ground Data Handling System (TV-GDHS) in the SFOF provides direct support to the SSAC group in the form of processed electrical video signals and finished photographic prints. The TV-GDHS operates as a service organization within the MOS structure. Documentation, system checkout, and quality control within the system are the responsibility of the TV-GDHS Operations Manager. During operations support the TV-GDHS Operations Manager reports to the SSAC Director.

4. Data Processing Personnel

The use of the Data Processing System (DPS) by *Surveyor* is under the direction of the Assistant Space Flight Operations Director (ASFOD) for Computer Programming. His job is to direct the use of the DPS from the viewpoint of the MOS. He communicates directly with the Data Chief, who is in direct charge of DPS personnel and equipment. Included among these personnel are the I/O console operators throughout the SFOF, as well as the equipment operators in the DPS and Telemetry Processing Station (TPS) areas.

Computer programs are the means of selecting and combining the extensive data processing capabilities of electronic computers. By means of electronic data processing, the vast quantities of mission-produced data are assembled, identified, categorized, processed and displayed in the various areas of the SFOF where the data are used. Their most significant service to the MOS is providing knowledge in real-time of the current state of the spacecraft throughout the entire mission. This service is particularly important to engineers and scientists of the technical support groups since, by use of the computer programs, they can select, organize, compare and process current-status data urgently needed to form their time-critical recommendations to the SFOD. (See Section V-C-3 for a description of the DPS.)

5. Other Personnel

The Communications Project Engineer (PE) controls the operational communications personnel and equipment within the SFOF, as well as the DSN/GCS lines to the DSIF stations throughout the world.

The Support PE is responsible for ensuring the availability of all SFOF support functions, including air conditioning and electric power; for monitoring the display of *Surveyor* information on the Mission Status Board and throughout the facility; for directing the handling, distribution, and storage of data being derived from the mission; and for ensuring that only those personnel necessary for mission operations are allowed to enter the operational areas.

The DSIF Operations Planning PE is in overall control of the DSIF Stations at Goldstone, Johannesburg, and Tidbinbilla; his post of duty is in the SFOF in Pasadena. At each station, there is a local DSIF station manager, who is in charge of all aspects of his DSIF station and its operation during a mission. The *Surveyor* personnel located at each station report to the station manager.

B. Mission-Dependent Equipment

The mission-dependent equipment consists of special hardware provided exclusively for the *Surveyor* Project to support the Mission Operations System. Most of the equipment in this category is contained in the Command and Data Handling Consoles and Spacecraft Television Ground Data Handling System which are described below.

1. Command and Data Handling Console

The Command and Data Handling Console (CDC) comprises that mission-dependent equipment, located at the participating Deep Space Stations (DSS), that is used to:

1. Generate commands for control of the *Surveyor* spacecraft.

2. Process and display telemetered spacecraft data.

3. Process, display, and record television pictures taken by the spacecraft.

At the DSIF installations, the CDC provides command, telemetry, and television functions through the DSIF antenna, transmitter, and receiver to the spacecraft. The CDC also provides telemetry data for on-site data processing (OSDP) at the DSIF stations and for relay to the SFOF at JPL for analysis.

Table VI-1. Mission-dependent equipment support of
Mission A at DSIF stations

DSS 11, Goldstone	Prime station with command, telemetry, and TV
DSS 42, Canberra	Prime station with command, telemetry, and TV
DSS 51, Johannesburg	Prime station with command and telemetry
DSS 61, Madrid	Station used for monitoring and training
DSS 72, Ascension	Station used for monitoring and training
DSS 12, Goldstone	Station configured for command backup
DSS 14, Goldstone	Station configured to record terminal descent, landing, and TV

Table VI-2. Mission A Command and TV activity

Station	Commands transmitted	TV frames received
DSS 11	89,778	10,980 (approx)
DSS 42	12,547	4182* (approx)
DSS 51	6780	712* (approx)
DSS 61	429	383*
DSS 72	0	**

*The majority of TV frames received by overseas stations were commanded
from DSS 11, Goldstone, during overlap coverage.
** No TV equipment.

a. *Network configuration.* Table VI-1 lists the mission-dependent equipment support of Mission A at the DSIF stations. CDC's were located at DSS 11, 42, 51, 61, and 72. Stations 11, 42, and 51 were the prime *Surveyor* stations. However, Station 61, at Madrid, was used for some command transmissions during the second lunar day. This station was primarily used for monitoring telemetry data and to provide station personnel with operational training. Likewise, DSS 72 at Ascension was used for monitoring and training. DSS 12, the Echo site at Goldstone, was configured and checked out to provide command backup to DSS 11. The Echo Station transmitter was set to the *Surveyor* frequency and a patchable interface established via the intersite microwave link for the command sub-carrier from the CDC at DSS 11. A return link was also established from the DSS 12 receiver back to the DSS 11 CDC for purposes of checking the command transmission. DSS 14, Mars Site at Goldstone, was configured to record the terminal descent and landing telemetry and post-landing TV data. Both pre- and post-detection signals were recorded on magnetic tape. The added capability of this station was used to increase the probability of obtaining touchdown data. Additionally, the VR-1560 tape recorder was used to record TV sequences during the first few days.

b. *CDC operations.* During the mission, CDC operations were conducted at five DSIF stations. Table VI-2 shows the number of commands transmitted and the number of television frames received by each station during the mission.

1. *DSS-11 Goldstone.* Pioneer Station, at Goldstone, participated in 17 passes during the first lunar day and 11 passes during the second lunar day. This station used a full CDC with command, telemetry,

and TV equipment. Three additional interfaces were established, as follows:

a. During telemetry sequences, the telemetry sub-carriers being received were transmitted to the SFOF from the CDC via the "96-kc" line. Signal-limiting and level adjustments were provided by the CDC.

b. As mentioned earlier, an interface was established with DSS 12, Echo Station, for command backup. If necessary, the CDC command SCO could be patched to the intersite microwave link for transmission to Echo Site, where the S-band transmitter would be modulated. A detected signal from the Station 12 transmitter was fed back to the Pioneer Station CDC via another microwave channel for checking command transmissions in the CDC.

c. Two dataphone links were established with Hanger AO at ETR. One line carried the reconstructed telemetry PCM waveform from the CDC's decommutator to ETR, and the other carried the command bit stream, obtained at the CDC system tester, to ETR. Both these lines were used during the first three view periods and then their use was discontinued.

2. *DSS 42 Canberra.* This station participated in sixteen passes during the first lunar day and 11 passes during the second lunar day. The CDC configuration was standard with full capability available. This station also recorded the voice experiment, "voice from the moon," and after being processed through some tunable filters, the recording was quite readable.

3. *DSS 51 Johannesburg.* This station participated in 17 passes during the first and 8 passes during the second lunar day. The CDC at DSS 51 was not officially committed for television for this mission. However, some television equipment tests had been run and the performance of the station was fairly well understood. This CDC lacks a spare 35mm photorecorder, and no video tape recorder was available. Despite this, some television activity was conducted by the station.

4. *DSS 61 Madrid.* The CDC at DSS 61 was used primarily for monitoring the spacecraft signal and training its crew. This station did not participate in pre-mission operational training tests. The station participated in three passes during the first lunar day and 5 passes during the second lunar day. At times when the data into SFOF from Station 51 were of poor quality, DSS 61 data, which were being sent to SFOF, were used. The CDC at DSS 61 was of standard configuration, except that the spare 35mm film recorder was lacking.

5. *DSS 72 Ascension.* The CDC at DSS 72 is limited to PCM telemetry operation and manual command transmission. No television or analog telemetry equipment is provided. This station acquired the spacecraft signal on the first pass and started processing telemetry data and transferring the data to SFOF via TTY lines. It also monitored the spacecraft acquisition by DSS 51. The station was officially secured

after this first pass, but decommutator lock on telemetry data was again obtained on June 2, and the station obtained two-way phase-lock on June 10.

c. CDC Mission A Anomalies. Table VI-3 lists the CDC equipment anomalies that occurred during the countdown and operations for Mission A. The corrective action stated is that action occurring up to this time; in some cases additional action is still required. Station 72 had no CDC equipment problems during Mission A.

2. Spacecraft Television Ground Data Handling System

The spacecraft Television Ground Data Handling System (TV-GDHS) was designed to record on film the television images received from *Surveyor* spacecraft. The principal guiding criterion was photometric and photogrammetric accuracy with negligible loss of information. The system was also designed to provide display information for the conduct of mission operations and for the production of user products such as prints, enlargements, duplicate negatives, and catalogs of ID information.

The system has been implemented in two parts: at DSS 11, Goldstone, and at the SFOF, Pasadena. At DSS 11 is an On-Site Data Recovery Subsystem (OSDR), and an On-Site Film Recorder Subsystem (OSFR). These subsystems are duplicated in the Media Conversion Data Recovery Subsystem (MCDR) and in the Media Conversion Film Recorder Subsystem (MCFR) at the SFOF. Also at the SFOF is the rest of the complete system.

Table VI-3. CDC Mission A anomalies

Pass	Anomaly	Effect on mission	Immediate reaction	Corrective action
		DSS 11		
2	Command buffer power-supply voltage read low	None	Replace power supply with spare	Bad solder joint resoldered and unit returned to service
5	Low-frequency oscillograph had noisy traces	None	Replace transistor on power-supply regulator card	No further action
3, 5, 6, 16	Command printer had intermittent sticking of 2 print wheels	None	When problem occurred, correct command verified by telemetry response and binary command comparator	Printers aligned, cleaned, and returned to service. Units returned to manufacturer for overhaul after first lunar sunset and returned for use during second lunar day
Several	Tape reader intermittently read wrong character and stopped on parity check	Occasional delay in TV sequence	Used spare tape reader	Unit adjusted and cold solder joint found. Units returned to manufacturer for overhaul after lunar sunset and returned for use during second lunar day

Table VI-3. (Cont'd)

Pass	Anomaly	Effect on mission	Immediate reaction	Corrective action
13	About 20% of TV ID frames not processed by decommutator	Loss of some TV ID data	Careful tuning of receiver and temporary circuit added to prevent video saturation	Investigation continuing to determine cause of problem
DSS 42				
4	TV monitor tube had excessive corona	None	Tube replaced with spare	No further action
10	Tape punch not advancing	None	Used spare tape punch	Unit repaired by replacing forward escapement mechanism
8	Telemetry display meter 16 occasionally sticks at 100%	None	Item usable despite problem	Meter was replaced
All	Command generator elapsed time counter wheel occasionally sticks	None	Item usable despite problem	Part was replaced
38	When Command 1115 was transmitted, command display changed to 1155. Command was transmitted correctly	None	Item usable despite problem	Circuit susceptible to noise. Circuit card replaced with newer type
DSS 51				
1	Command 0316 not transmitted correctly	None	Command was retransmitted correctly	Problem found to be due to operator shorting command signal at monitor point. Stations were advised of operational procedure to prevent shorting signal. Investigating nonshorting type jacks, signal monitoring at reproduce module output and additional isolation
6	Decommutator gave excessive parity errors	None	Transferred to spare decommutator	Problem due to dirty switch contact. Switch cleaned and unit returned to use
39	Command generator sequence counter would not reset to zero	None	Item usable despite problem	The mechanical assembly was adjusted and item returned to use
42	Decommutator did not up-date local telemetry display	None	Transferred to spare decommutator	Three transistors were replaced and unit returned to use
DSS 61				
44	TV photorecorder clock not operating	None	Replace battery	Same
44	TV generator output faulty in 200-line mode	None	Use as is	Unit replaced with one of new design

a. Equipment at Pioneer Site, Goldstone (TV-11). Data for the TV-GDHS is injected into the system at the interface between the Station 11-receiver and the OSDR. At this point, the signal from *Surveyor* has been down converted to a 10-Mc FM-modulated signal. The OSDR further down converts the signal to 5 Mc (500 kc for 200-line mode), providing an output signal for use by the station's FR-800 videotape recorder and to the Microwave Communication Link for transmission to the SFOF. A baseband signal (the 500-kc predetection signal for 200-

line mode) is provided to the station's FR-1400 instrumentation tape recorder. The OSDR further processes this signal to obtain television image synchronization, telemetry synchronization, and the base-band video signal. This information is then used by the OSFR to record the video image and the raw ID telemetry in bit form on 70mm film, together with an internally generated electrical gray scale and "human readable" time and record number.

The entire on-site portion of the TV-GDHS at DSS 11, Goldstone, was committed for Mission A for both 200- and 600-line modes of operation. Approximately 4,000 ft of 70mm film was exposed during the first lunar day, producing some 10,300 frames of near-perfect film recordings of the *Surveyor* spacecraft images. During the first lunar day, only 26 frames are known not to have been recorded. There was no known loss of lines in any of the pictures recorded; jitter was less than ½ pixel (picture element), almost unmeasurable, and resolution appears to be as good as was received from the spacecraft. In all frames, a very light vertical stripe appears in the black portions of the image which has been traced to a grounding problem in the TV-synchronizer. Recent evaluation of the films has revealed a slight shading from one corner of the image to the opposite corner. System gamma, which can be defined for any point in the system, is here used as the relationship of the input voltage to the density on the negative after development. The desired system gamma was 1. A spot check of the actual system gamma indicates that it lies between 1.06 and 1.16. TV-11 was able to receive and lock on to the very first picture transmitted by the spacecraft. In 200-line mode, the frequency uncertainty of the signal from the spacecraft is 8 times the bandwidth of the signal. Considerable prior effort was expended by personnel from TV-11, TV-1, DSS 11, and SSAC to develop the procedures necessary to make possible the recording of this first picture. TV-11 performed the fine tuning for both TV-11 and both portions of TV-1 in 200-line mode.

b. Equipment at the SFOF (TV-1). The signal presented to the Microwave Terminal at DSS 11 is transmitted to the SFOF where it is distributed to the MCDR. The MCDR processes the signal in the same manner as the OSDR. An FR-700 video magnetic tape recorder records the predetection signal in the same manner as the FR-800 at DSS 11. In addition, the MCDR passes the raw ID information to the Media Conversion Computer (MC Computer), which converts the data to engineering units. This converted data is passed to (1) the film recorder, where it is recorded as "human readable" ID, (2) the ID wall display board in the SSAC area, (3) the disc file, where the film chip index file is kept, and (4) the history tape.

Because of implementation constraints, a preliminary configuration consisting of a so-called film recorder simulator and a breadboard data recovery unit was provided to drive the scan converter. This configuration together with the FR-700 tape recorder was committed for Mission A.

The scan converter accepts the slow scan image information from the film recorder simulator and converts it to the standard RETMA television signal for use by the SFOF closed-circuit television and the public TV broadcast stations.

The MC film recorder provided two films. One of these films is passed directly to the bimat processor; the other is accumulated in a magazine. This film is wet-processed off-line.

The bimat processor laminates the exposed film with the bimat imbibed material, producing a developed negative and a positive transparency. The negative is used to make strip contact prints, which are delivered to the users. The negative is then cut into chips and entered into the chip file, where they are available for use in making additional contact prints and enlargements.

The additional equipment in the SFOF was used on a best-effort, noninterference basis during Mission A. It was not committed to the mission because there was insufficient time after installation to properly develop the operational procedures and to properly train the manning personnel. The only operational tests in which there was any participation by the manning personnel were the two "C" tests prior to the mission.

Performance of the "committed" equipment in the SFOF. The committed portion of the TV-1 at the SFOF performed nominally throughout the mission. All 200-line-mode pictures were received in their entirety. In 600-line mode, about a dozen pictures were lost because of spacecraft signal drift near the end of the first lunar day. The FR-700 video magnetic tape recorder made very good predetection recordings. However, tapes recorded prior to June 11, 1966, have a lower-than-nominal time code recording, making it difficult but not impossible to recover time. It was found that an improper signal level was being fed to the recorder because of operator error.

Performance of the additional equipment in the SFOF. Approximately 33,000 ft of 70mm film was exposed during the entire mission in the MCFR, using both Optics No. 1 and No. 2. When the bimat processor is run at its maximum speed, a finished strip contact print can be produced within 30 min of a given exposure if the film strip is no more than 50 frames long. However, this is quite wasteful of film. After the second or third video pass, it was realized that there was no real need for such speedy processing. The real need for quick prints was supplied by the SSAC paper camera, which produces paper prints within 10 sec after exposure. These prints are of adequate quality for the purpose of real-time mosaicking. Therefore, a slower mode of operation was used, saving several thousand feet of film. Toward the end of the first lunar day, 4000 ft of film was borrowed to ensure the film supply, since more pictures were being taken than previous estimates had anticipated. The bimat processor used approximately 21,000 ft of bimat material during the entire mission in developing the film from Optics No. 1. The images recorded on Optics No. 1 were very good throughout most of the mission.

On June 3, 1966, some scan lines were missed and the problem was traced partially to the TV synchronizer. It was later found, after more serious line dropping occurred, that there was a defective isolation amplifier in the Communications Center.

No film recording on either Optics was made on June 10 because of several component failures in the film recorder. After all problems were resolved, it was discovered that the cathode ray tubes (CRTs) in both Optics have slight horizontal burns near the top of the image area. The image area was moved downward slightly to remove it from the burned area. The June 10 data were later recorded on film, using the magnetic tape recording from the FR-700. On June 11, a failure in Optics No. 2 caused serious burns on the face of the CRT in Optics No. 2. Optics No. 1 was not affected. The images recorded on Optics No. 2 prior to June 4 were found to be slightly out of focus and were over-dithered. Vertical dithering of the recording beam is used to reduce the apparent line structure of the image, and does not impair horizontal resolution. Too much dither causes line overlap and vertical resolution will be impaired. After replacement of the CRT of Optics No. 2, the film produced appeared to be very good.

A spot check of the system gamma on Optics No. 2 shows it to be about 0.96. Approximately 14,000 ft of strip printing paper was used in producing the strip contact prints for the Science Advisory Team and the mosaickers.

The strip contact prints were of very good quality and were extensively used by the Science Advisory Team and the mosaickers. Spherical mosaics were produced with these prints. Several flat mosaics were made. The printing paper used was especially selected for mosaicking purposes.

All enlargements and individual chip contact prints were made using the Ektomatic process and the variable-contrast Ektomatic paper. The quality achieved with this paper was consistent with the capabilities of the paper. Some 2000 8 × 10 enlargements of selected film chips were provided during the first lunar day, and an average of approximately 140 enlargements per day have been supplied since then.

A film chip catalog has been produced. The catalog consists of the spacecraft ID as received by the computer, mission-independent parameters such as time, derived parameters such as the rotation angle, and computer-derived evaluation of the ID quality and alarms. When the chip index file has been manually updated, the catalog then produced will also contain the manual inputs, such as corrected ID and searchable and nonsearchable comments. It has been estimated that roughly 5% of the catalog contains data that were suspect for one reason or another.

The search program was very useful to the Science Advisory Team and was highly praised by them, even though it is known that many improvements can be made in the program to facilitate its use.

C. Mission Operations Chronology

Inasmuch as mission operations functions were carried out on an essentially continuous basis throughout Mission A, only the more significant and special, or non-standard, operations are described in this chronicle. Refer also to the sequence of Mission A events presented in Table A-1 of Appendix A.

1. Countdown and Launch Phase

Early in the countdown the *Surveyor* mission control interface was established between the SFOF, Pasadena, and AFETR at Cape Kennedy. The mission control interface consists of a voice communications circuit over which the launch operations and space-flight operations countdowns are communicated. During the countdown, AFETR reported problems with the following: Antigua radar equipment; the communications link with the Range Instrumentation Ship (RIS) *General Arnold;* and readiness of the *General Arnold* to provide tracking and telemetry coverage. During the countdown and launch, temporary failures also occurred in the DSS 51 high-speed data and voice communications circuit. Mission Control advised the Mission Director that the loss of tracking coverage from the *General Arnold*, although undesirable, should not delay or postpone the launch. These problems are discussed in greater detail in Sections II and V.

Launch occurred at 14:41:00.990 GMT on May 30, 1966. A description of the sequence of events from launch through separation is contained in Section III-D. Launch vehicle telemetry data confirmed that the *Centaur* issued the preprogrammed commands to the spacecraft before separation. These were required to extend the landing legs, extend the two omnidirectional antennas, and switch the transmitter from low to high power. Spacecraft telemetry indicated proper response to these signals except that extension of Omniantenna A could not be confirmed.

Following separation at 14:53:38 GMT, the spacecraft executed the planned automatic sequences as follows. By using its cold-gas jets, which were enabled at separation, the flight control subsystem nulled out the small rotational rates imparted by the separation springs, and initiated a roll-yaw sequence to acquire the sun. After a minus roll of approximately 100.5 deg and a plus yaw of 86 deg, acquisition and lock-on to the sun by the spacecraft sun sensors were completed at 15:00:34 GMT. Concurrently with the sun acquisition sequence, solar panel deployment was initiated, and at 15:03:20 GMT the solar panel was in its proper transit position. All these operations were confirmed in real-time by telemetry received via downrange ships.

Following sun lock-on, the spacecraft coasted with its pitch and yaw axes controlled to track the sun and with its roll axis held inertially fixed.

2. DSIF and Canopus Acquisition Phase

One-way communication (down-link) with the spacecraft was achieved within 11 sec of first predicted visibility to DSS 51. In less than 4 min, two-way communication was established to complete acquisition at 15:09:00 GMT. It was later reported that DSS 51 tracking data looked good. It was also reported that spacecraft telemetry data looked good, except that Omniantenna A extension was not indicated.

The initial sequence of commands was sent to the spacecraft, beginning at 15:20:59. This sequence consisted of commands to (1) turn off spacecraft equipment required only until DSIF acquisition, such as high-power transmitter and accelerometer amplifiers, (2) lock the solar panel position, (3) increase the telemetry sampling rate to 1100 bit/sec, and (4) perform the initial interrogation of all telemetry commutator modes. Spacecraft response to these commands was normal. However, telemetry continued to indicate that Omniantenna A was not extended. A nonstandard sequence of commands to ignite the omniantenna release squib was initiated to attempt to extend Omniantenna A. These commands were initiated at 16:21:21 GMT with negative results. A preliminary analysis of the effect of the antenna anomaly on mission performance was conducted by the SPAC and FPAC groups. The results of this study indicated the feasibility of designing a midcourse maneuver using Omniantenna B.

SPAC recommended to Mission Control that the Canopus acquisition sequence be performed early because (1) indications were that the Canopus intensity signal was a little higher than expected for "no star in the field of view," and (2) a possibility existed that verification of the state of Omniantenna A extension might be obtained during the spacecraft roll. It was desired that satisfactory operation of the Canopus sensor and the state of Omniantenna A extension be verified so that planning for the midcourse correction could be conducted accordingly.

At 18:50:44 GMT, commands were sent to the spacecraft to maintain sun lock and initiate a roll attitude maneuver for the Canopus acquisition sequence. Star intensity signals from seven stars in addition to Canopus were identified during the first revolution, with Canopus being observed and identified after 220 deg of roll. As the star sensor swept past Canopus, the star intensity level indicated saturation and the Canopus lock-on signal did not appear. From the 360-deg roll maneuver data, it was concluded that manual lock-on would have to be used to acquire Canopus. As the spacecraft continued to roll, the time at which to stop the vehicle roll (so that Canopus would be within the field of view of the star sensor) was computed. At 19:13:20 GMT, a manual lock-on command was sent to the spacecraft which successfully terminated the roll maneuver and positioned the star sensor on Canopus. The spacecraft then continued to coast as before, but with its roll attitude controlled so that the Canopus sensor remained locked on the star.

During the Canopus acquisition roll maneuver, the Receiver A AGC data were monitored so that information could be obtained which might verify Omniantenna A nonextension. The AGC data appeared to confirm that

Omniantenna A was not extended when compared with data from antenna radiating pattern tests, which were run concurrently using a spacecraft model on the ground. However, the results were not conclusive.

3. Premidcourse Coast Phase

At 20:54:02 GMT, commands were sent to the spacecraft to initiate gyro drift checks which lasted about 3 hr. This was done approximately 1 hr earlier than originally planned in order to obtain as much drift data as possible prior to the midcourse maneuver. The drift rates were found to be within specifications.

In preparation for the first of two premidcourse maneuver conferences, SPAC prepared an evaluation of spacecraft performance and FPAC computed a preliminary orbit determination and assessed the overall possibilities of getting the spacecraft to the preselected landing site. The first premidcourse maneuver conference was held about 7 hr after launch. The spacecraft performance and subsystem margins as well as the midcourse maneuver possibilities were presented in a general status review to the Mission Director.

During this low-power coast phase, continuous telemetry was received at 1100 bit/sec until approximately 11 hr after launch and at 550 bit/sec until midcourse maneuver preparations. Three engineering interrogations were performed, including two originally scheduled for transmitter high power. All systems were performing satisfactorily.

In preparation for the second premidcourse maneuver conference, SPAC continued to assess spacecraft performance and FPAC continued monitoring tracking data. After FPAC completed an intermediate midcourse maneuver computation, the FPAC and SPAC groups convened to analyze midcourse maneuver alternatives and considerations. Next, a preliminary internal mission operations conference was held between the Space Flight Operations Director (SFOD) and the SPAC and FPAC group representatives to generate a list of alternate midcourse maneuver plans and prime selected landing sites for presentation to the Mission Director.

The second premidcourse maneuver conference was held about 12 hr after launch. The Mission Director approved the following midcourse maneuver plan:

1. The midcourse maneuver to be executed approximately 15.85 hr after injection.

2. A roll-yaw premidcourse attitude maneuver sequence to be utilized.

3. Omniantenna B to be used with Transponder B in two-way lock.

4. Spacecraft transmitter to be in high-power mode with a telemetry sampling rate of 4400 bit/sec.

5. The velocity correction to be approximately 20 m/sec.

6. The aiming point to be moved slightly to −2.33 deg latitude, 316.17 deg longitude.

7. The coast mode commutator to be used for the post-midcourse reverse maneuvers to obtain more Receiver A AGC data for further analysis of Omniantenna A status.

(Refer to Section VII-C for a discussion of the factors which were considered in selecting the midcourse maneuver magnitude and final aiming point.)

The SFOD's maneuver plan was based upon orbit computations using the tracking data obtained prior to the first DSS 11 view period. (See Appendix A, Table A-2, for predicted view period summary.) After DSS 11 had tracked the spacecraft for a time sufficient to obtain the required number of data points, the final midcourse maneuver parameters were computed.

4. Midcourse Maneuver Phase

The spacecraft was prepared for the midcourse maneuver by sending commands to (1) turn on the transmitter high power, (2) increase the telemetry sampling rate from 550 to 4400 bit/sec, (3) pressurize the vernier propulsion system, and (4) null out Canopus error signals. The commands necessary for these preparations were transmitted within the 20-min period preceding the first premidcourse attitude maneuver. Starting at 06:30:13 GMT on May 31, the required roll (−86.5 deg) and yaw (−57.9 deg) attitude maneuvers were commanded and executed to align the vernier engine axes in the direction of the desired midcourse velocity vector correction. Then, at 06:45:03 GMT, the command was sent which started the vernier engines for the velocity correction. Based upon a time increment of 20.8 sec, previously commanded and stored by the spacecraft flight control system, midcourse thrust was terminated automatically by the spacecraft after an actual duration of 20.75 sec.

Following midcourse thrusting, commands were sent to reduce the telemetry bit rate (to observe Receiver A AGC) and execute the reverse attitude maneuvers. At completion of the reverse maneuvers, the sun and Canopus were in the field of view of the sensors, and lock-on was achieved without the necessity of performing a roll maneuver for star verification. Next, a high-power engineering interrogation was conducted, after which the spacecraft was commanded to transmitter low-power operation at 07:12:43 GMT.

5. Post-Midcourse Coast Phase

Following midcourse, the SPAC director recommended that the spacecraft remain in the inertial mode in roll. This was requested in order to obtain more roll gyro drift data and to obtain data on the decrease of the Canopus intensity signal as a function of the angle off of the sensor axis. Four additional 3-hr gyro drift checks were conducted during the post-midcourse coast phase, the results of which were consistent and indicated gyro drift was below the specified limits. The gyro drift data was given to FPAC for terminal descent computations.

A second attempt was made to command the extension of Omniantenna A, although analysis indicated that sufficient margin existed relative to the effects of the center-of-gravity offset on spacecraft stability during terminal descent. Again, the results were negative.

Also during this phase, power-mode cycling checks were made to help predict the percentage of the electrical load that would be supplied by each of the spacecraft batteries (main and auxiliary) during the terminal descent when both batteries are placed directly on the power buss. Following these checks, a slight change was made in the procedure in order to place both batteries directly on the line. The auxiliary battery was used to supply most of the load so that its temperature would be increased to the desired temperature of >60°F at the time of terminal descent.

The spacecraft was also commanded to turn on heaters for thermal control of the vernier system propellant tanks.

Telemetry signal levels were sufficient to permit the telemetry sampling rate to be maintained at 550 bit/sec throughout this coast phase, during which time 10 low-power engineering interrogations were made.

6. Terminal Maneuver and Descent Phase

In preparation of the first terminal maneuver conference, the SFOD held a preliminary internal mission operations conference to generate a list of alternate terminal maneuver plans and an outline of post-landing operations for presentation to the Mission Director.

The first terminal maneuver conference was held about 15 hr before estimated main retro ignition. The Mission Director decided which of the alternatives were to be considered for further investigation and analysis. The specific terminal maneuver tentatively selected by the Mission Director consisted of roll-yaw-roll maneuvers using Omniantenna B. These maneuvers optimized the signal strength of Omniantenna B, were the best choice for the Project Scientist (inasmuch as the surface unobscured by the spacecraft would be toward the sun, providing good contrast) and used the attitude maneuvers (roll and yaw) already proven in conducting the midcourse correction.

The second terminal conference was held about 5½ hr before main retro ignition. The SFOD presentation covered four items of consideration:

1. No new data had been received to invalidate the previously selected terminal maneuver.

2. A final decision must be made between a roll-yaw-roll maneuver (which was tentatively selected) and a roll-pitch maneuver (which also optimized Omniantenna B signal strength but was less desirable).

3. A decision was required as to whether the strain gage power should be turned on to obtain touchdown dynamics data, inasmuch as there was concern that the communications signal margins would not permit normal telemetry plus the strain gage at touchdown.

4. A decision was required to select a post-touchdown operations plan relative to the choice of 200- or 600-line video picture sequences.

The Mission Director approved the following terminal maneuver plan:

1. The roll-yaw-roll maneuver would be used.

2. The touchdown dynamics strain-gage power would be turned on as there appeared to be a sufficient signal margin.

3. The post-landing operations sequence employing the 600-line video mode would be employed after the initial 200-line sequence.

(Refer to Section VII-D for additional discussion of the terminal maneuver analysis.)

NASA/JPL TECHNICAL REPORT NO. 32-1023

129

About 5 hr before main retro ignition, in preparation for terminal descent, the spacecraft was commanded to switch to the main battery mode, in which both batteries are on-line. The last roll gyro drift check was terminated 28 min later, and the accumulated error was nulled.

About an hour before retro ignition, transmitter high power was turned off and the last engineering interrogation was performed. In addition, the telemetry rate was increased to 1100 bit/sec and power was turned on as required for monitoring vernier propulsion and touchdown forces by means of strain gages. The spacecraft transponder was then turned off to put DSS 11 on one-way lock for the terminal maneuvers.

The terminal roll-yaw-roll maneuvers were commanded 38, 33, and 29 min before retro ignition, respectively. These were: roll +89.3 deg, yaw +59.9 deg, and roll +94.1 deg. The first roll and yaw maneuver oriented the retro thrust axis, and the final roll oriented the spacecraft for optimum telecommunications from Omniantenna B during terminal descent and optimum camera and sun relationship for photographs after landing.

In the remaining time before the automatic retro sequence, final preparations were made for the terminal descent. The thrust level of the vernier engines was preset to a nominal 200-lb total, and a delay time of 7.826 sec between the "AMR mark" event and ignition of the vernier engines was stored in the spacecraft. The altitude marking radar (AMR) was turned on and then enabled less than 2 min before retro ignition. Except for two commands for turning on the touchdown strain-gage data and obtaining the optimum sampling rate for the flight control telemetry data, the remaining operations were automatically performed by the spacecraft.

The terminal descent sequence, initiated by the AMR "mark" signal, occurred at 06:14:40.7 GMT, June 2, and proceeded automatically through touchdown. (A backup AMR "mark" time was also commanded.) The descent sequence included the following events: vernier ignition, main retrorocket ignition, RADVS turn-on, retrorocket burnout and separation, doppler control enabling, and vernier engines shutoff. Touchdown occurred at 06:17:37 GMT ±0.5 sec on June 2, 1966 at −2.58 deg latitude and 316.65 deg longitude.

A summary of the significant deviations from the standard Mission A flight sequence is given in Table VI-4.

Table VI-4. Deviations from standard Mission A flight sequence

1. Implementation of nonstandard sequences in an attempt to extend Omniantenna A, which had failed to extend normally prior to Centaur/Surveyor separation.

2. Performance of all high-power interrogations (except that conducted as part of the terminal descent sequence) in low power because of sufficient telecommunication signal strength margin.

3. Use of manual lock-on (instead of automatic lock-on) for Canopus acquisition because of high signal level produced by all stars, including Canopus.

4. Additional gyro drift checks obtained during postmidcourse coast phase to obtain best estimate of gyro drift for use in terminal descent computations.

5. Additional use of auxiliary battery to increase the auxiliary battery temperature to that desired during terminal descent.

6. Three attitude maneuvers (instead of two) used during terminal descent.

7. Lunar Phase

In the first few minutes after touchdown, engineering interrogations indicated the spacecraft had survived the landing and was in excellent condition. Telemetry indicated Omniantenna A was now in the fully extended position, apparently having deployed as a result of retro or touchdown shock.

A nonstandard sequence of commands was sent about 10 min after touchdown, turning off the AMR heater power to minimize battery power consumption. Thirteen min later the spacecraft was commanded to connect the main and auxiliary batteries, in parallel, through isolation diodes to the unregulated power buss, and remove the direct connection between the auxiliary battery and the unregulated power buss.

During the next several minutes, commands were sent to prepare for and take the first TV pictures in 200-scanline, wide-angle (25 deg field of view) mode. The survey camera iris servo loop was first closed so that the f-stop would be controlled automatically. The first picture was received at 06:53 GMT, showing Footpad 3 and a portion of the spacecraft against the surface of the Moon. During the next 51 min, 13 additional 200-line pictures were taken in a survey between Footpads 3 and 2.

After the series of 200-line pictures had been obtained, the spacecraft was commanded to an engineering telemetry configuration and interrogated to assess the performance of the subsystems. This completed the standard command sequences.

At the second post-midcourse maneuver conference, the decision had been made to reconfigure the spacecraft and DSN Ground Communications System for 600-line pictures after the first 200-line sequence was completed. Accordingly, one of the prepared optional command sequences was utilized to accomplish the following: (1) erection of the solar panel, (2) disconnection of the auxiliary battery from the unregulated power buss, (3) erection of the high-gain antenna, and (4) preparation for 600-line pictures.

The first 600-line, high-resolution picture sequence was initiated at 09:43 GMT, and 133 600-line pictures were taken before DSS 11 visibility ended on June 2, 1966. The 200-line configuration was not utilized again in video operations except for a few pictures taken by DSS 42 on June 5 and by DSS 51 at the end of the second lunar day on July 14.

The lunar operations sequences, other than routine engineering interrogations, are listed in Table A-3 of Appendix A. The lunar sequences consisted largely of picture surveys, which were interrupted by configuration changes for engineering interrogations. The vast majority of pictures were obtained using DSS 11, which is the only DSN station furnished with the special mission-dependent TV ground data handling equipment. This equipment is described in Section VI-B. A few pictures were obtained by DSS 51 and DSS 42 during the visibility periods of June 4 and 5, 1966, and again on the second lunar day.

Extensive horizontal- and vertical-scan picture surveys were conducted of the entire lunar surface and horizon visible to the *Surveyor* spacecraft. Use was made of both the wide- and narrow-angle zoom lens positions, as well as the red, green, and blue filters for color data. Many picture surveys were repeated to obtain comparisons at different sun elevation angles or under spacecraft shadow conditions. Surface pictures included numerous studies of imprints made by the spacecraft landing-leg footpads and crushable blocks.

It was possible to continue operation of the spacecraft over a much greater portion of the lunar day than had been expected. It had been anticipated that excessive temperatures could occur in critical systems if the spacecraft was operated during several days near lunar noon. However, by taking advantage of shading afforded by the solar panel and high-gain antenna, it was possible to perform operations on all but two days, June 8 and 9.

Experiments were conducted to observe possible lunar surface reaction to nitrogen gas jet impingement. In these experiments, photographs were taken before, during, and after opening the valve of an attitude-control gas jet which pointed toward the lunar surface.

Many camera surveys were made of the spacecraft itself to aid in engineering evaluation. Video data confirmed the post-touchdown extended position of Omni-antenna A, as well as the excellent condition of the spacecraft in general. Various spacecraft surfaces were viewed to determine possible color deterioration, dust coverings, or structural damage.

Several video sequences were performed to photograph bright stars and a portion of the earth to provide data for improving the estimated landing site location. The attempt to photograph the earth was unsuccessful owing to the limited elevation angle range of the camera mirror. Sequences were also obtained of the solor corona after sunset of the first lunar day.

After approximately 6000 video frames, a problem developed with the camera elevation potentiometer causing a half-step error in the readings. Later, on June 12, the potentiometer failed and elevation identification was no longer available. Modified operation was then initiated with minimum elevation stepping until it could be ascertained that it was safe to exercise the elevation drive at the previous high-duty cycle. The following day the failure was diagnosed as a break in the potentiometer wire which was bridged at a 17-deg mirror position, providing an elevation reading only at that angle. The nature of the failure was not considered critical. Consequently, normal elevation stepping was resumed.

Some difficulty also was encountered with the camera mirror elevation drive due to reduced torque as temperatures declined toward the end of the first lunar day. Desired elevation positions were attained by sending repetitive mirror step commands. This difficulty also was experienced at the close of the second lunar day.

As a result of the continued excellent status of the spacecraft as the first lunar day progressed, operations plans for the balance of the first lunar day were designed to permit the spacecraft to pass into the environment of the first lunar night with the highest possible chance of survival. Nearly a full charge was placed in the battery. The solar panel was positioned at 10 deg from the vertical, slightly upward toward the western lunar sky, to prevent a sudden surge of power to the battery before it could warm up during the second lunar day. The high-gain antenna was also positioned nearly vertical so that long shadows would be cast which could be photographed by the *Lunar Orbiter* spacecraft.

After sunset of the first lunar day, a final picture was taken at 15:37 GMT, on June 14. This was a 4-min exposure of Leg 2, illuminated by reflected light from the earth. Engineering interrogations were continued on July 15 and 16 to observe spacecraft performance when exposed to the severe thermal gradients. A final command was sent to the spacecraft at 20:31 GMT, on June 16, to cease telemetry transmission. Only the two receivers and command decoder were left functioning. Continued operation of these elements was necessary to enable the spacecraft to receive commands when attempts would be made to resume operations during the second lunar day.

On June 28, just a few hours before sunrise of the second lunar day, commands were sent from Goldstone to turn on the *Surveyor* transmitter. No response was received. For several hours each day thereafter, Goldstone repeated these commands. Beginning July 3, the task of repeating these commands was transferred to the Canberra station. Finally, at 11:29 GMT on July 6, *Surveyor* responded, and it was observed that battery voltage was very low. Cautious steps were taken to command the spacecraft to a configuration which permitted recharging of the battery.

Preparations were made to operate the camera on July 7, and 24 pictures were received which proved the camera to be in good working condition. Plans were then made to use the camera during the remaining three days before sunset. A battery anomaly was detected on July 8 when battery temperature began to rise rapidly. The solar panel was repositioned to reduce the charging rate and resultant internal heating, but the temperature continued to climb.

Emergency measures were then taken during the Canberra view period to utilize to best advantage what was then believed to be the last few hours of the spacecraft's operational life. Thirty-eight 600-line pictures were taken by DSS 42. Commands were sent to fire the three downward-pointing vernier engines at a low thrust level; but the attempt was unsuccessful because the power circuits which operate the propellant valves would not respond to turn-on commands.

Commands were also sent from DSS 42 to turn on the approach television camera, which had been designed for but not used during the terminal descent phase of flight. No video data was received, but telemetry indicated that camera turn-on was normal.

Battery temperature climbed to a peak of 139.8°F on July 8, but then began dropping and was within the normal operating temperature range by the following day. When the battery temperature had decreased sufficiently, the solar panel was repositioned in steps to increase the charge rate gradually until full charging rate was again achieved.

Though the battery anomaly was unresolved, the problem was less critical and camera operation was resumed July 12 via Goldstone. Among the pictures received that day was a survey of the top of spacecraft Compartment A, which revealed, for the first time, damage to the glass thermal control surface. As the sun sank lower in the sky, additional pictures were taken by all three prime DSIF stations to observe the spacecraft shadow as it advanced across the lunar landscape. Before sunset, the high-gain antenna was again positioned vertically, requiring reconfiguration to the 200-line mode for the final pictures. As the final pictures were taken by DSS 51, just prior to sunset in an attempt to photograph the solar corona, spacecraft telemetry indicated large battery voltage drops. On July 14, radio lock was lost by DSS 51 about 03:00 GMT, before the solar corona pictures could be commanded. Attempts were made to restore communications with the spacecraft, but the effort was abandoned when the DSS 51 view period ended.

The operational phase of Mission A was terminated at the conclusion of operations on July 14, 1966.

VII. FLIGHT PATH AND EVENTS

For Mission A, the landing site selected prior to launch for targeting of the launch vehicle ascent trajectory was near the western end of the *Apollo* landing zone at −2.58 deg latitude, 316.65 deg longitude (best estimate based on post-touchdown tracking data). The following factors influenced the selection of this site: predicted terrain smoothness, desire to land in the *Apollo* zone, desire to minimize the off-vertical incidence of the approach trajectory, and availability of good post-landing lighting. An unbraked impact speed was selected so that the Goldstone arrival visibility constraints would be satisfied for all launch days in the launch period.

A. Launch Phase

Liftoff occurred at 14:41:00.990 GMT, May 30, 1966, from ETR Launch Site 36A at Cape Kennedy, Florida. At 2 sec after liftoff, the *Atlas/Centaur* launch vehicle began a 13-sec programmed roll that oriented the vehicle from a pad-aligned azimuth of 105 deg to a launch azimuth of 102.285 deg. At 15 sec, a programmed pitch maneuver was initiated. The nominal and actual times for the boost phase and following mission events are summarized in Table A-1 of Appendix A. All event times were nominal or within the 3-σ tolerance. The launch-phase ascent trajectory profile is illustrated in Fig. VII-1.

Injection accuracy is discussed in Section VII-C in connection with midcourse correction requirements.

B. Cruise Phase

Separation of *Surveyor* from *Centaur* occurred at 14:53:37.9 GMT at a geocentric latitude and longitude of 17.6 and 312.1 deg, respectively. The spacecraft was in the sunlight at separation and never entered the shadow of either earth or moon during the transit trajectory. Figures VII-2 and VII-3 illustrate the transfer trajectory.

First acquisition of the *Surveyor* spacecraft by DSS 51 was both nominal and very quick. The $L-5$ min predictions indicated a rise over the station's horizon mask at 15:04:39 GMT. DSS 51 reported good one-way doppler data at 15:04:51 GMT and good two-way data at 15:08:31 GMT. Thus, DSS 51 was able to obtain good one-way data only seconds after spacecraft rise and good two-way data in less than 4 min after rise. Table A-2 of Appendix A presents the predicted view periods.

The proximity of the uncorrected, unbraked impact point (−11.425 deg latitude, 305.853 deg longitude) and the original aim point (−3.25 deg latitude, 316.17 deg

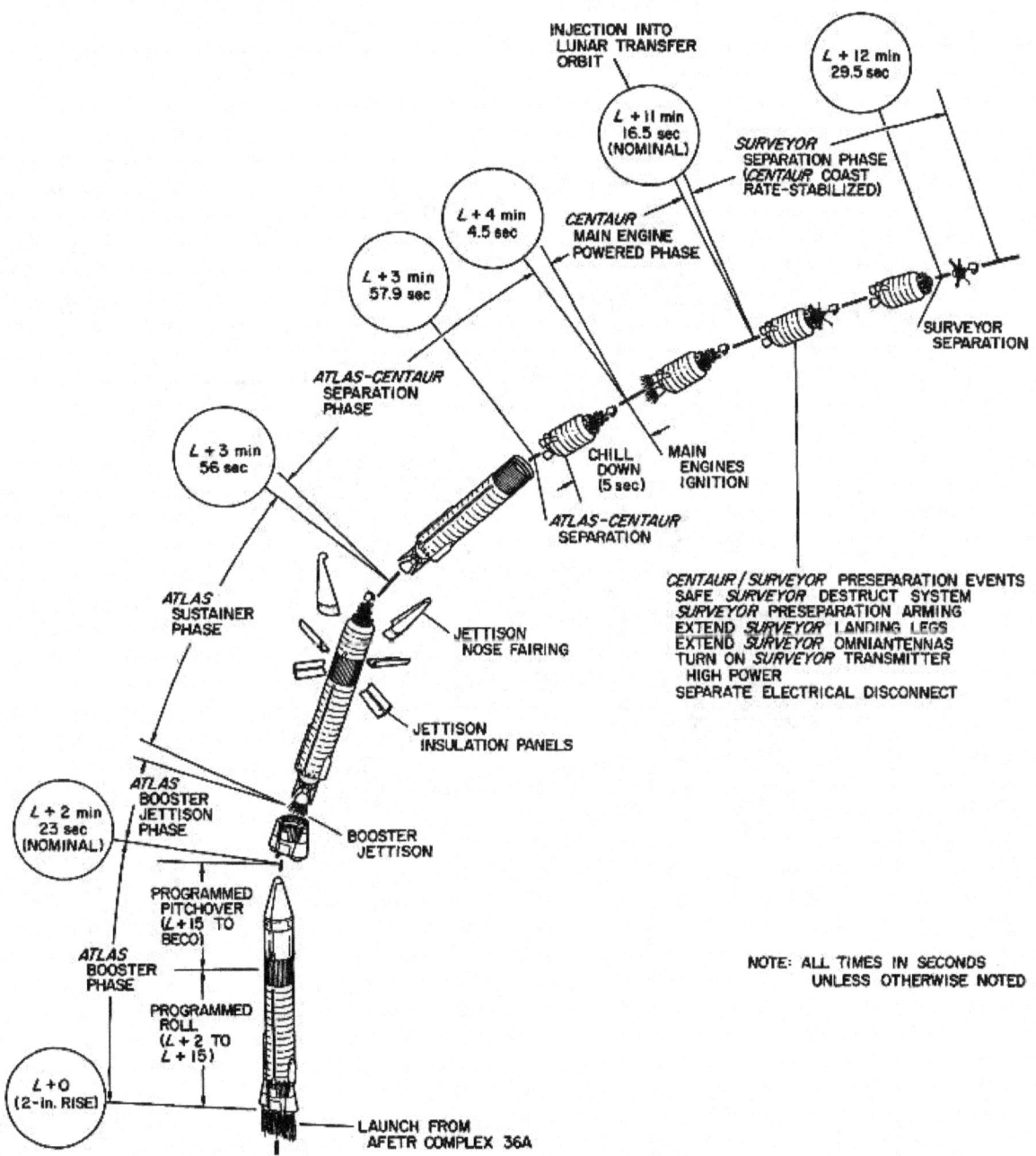

INJECTION INTO
LUNAR TRANSFER
ORBIT

L + 12 min
29.5 sec

L + 11 min
16.5 sec
(NOMINAL)

SURVEYOR
SEPARATION PHASE
(CENTAUR COAST
RATE-STABILIZED)

L + 4 min
4.5 sec

CENTAUR
MAIN ENGINE
POWERED PHASE

L + 3 min
57.9 sec

SURVEYOR
SEPARATION

ATLAS-CENTAUR
SEPARATION
PHASE

L + 3 min
56 sec

CHILL
DOWN
(5 sec)

MAIN
ENGINES
IGNITION

ATLAS
SUSTAINER
PHASE

ATLAS-CENTAUR
SEPARATION

JETTISON
NOSE FAIRING

CENTAUR/SURVEYOR PRESEPARATION EVENTS
SAFE SURVEYOR DESTRUCT SYSTEM
SURVEYOR PRESEPARATION ARMING
EXTEND SURVEYOR LANDING LEGS
EXTEND SURVEYOR OMNIANTENNAS
TURN ON SURVEYOR TRANSMITTER
    HIGH POWER
SEPARATE ELECTRICAL DISCONNECT

JETTISON
INSULATION PANELS

ATLAS
BOOSTER
JETTISON
PHASE

L + 2 min
23 sec
(NOMINAL)

NOTE: ALL TIMES IN SECONDS
    UNLESS OTHERWISE NOTED

BOOSTER
JETTISON

PROGRAMMED
PITCHOVER
(L+15 TO
BECO)

ATLAS
BOOSTER
PHASE

PROGRAMMED
ROLL
(L+2 TO
L+15)

L+0
(2-in. RISE)

LAUNCH FROM
AFETR COMPLEX 36A

Fig. VII-1. Launch phase trajectory

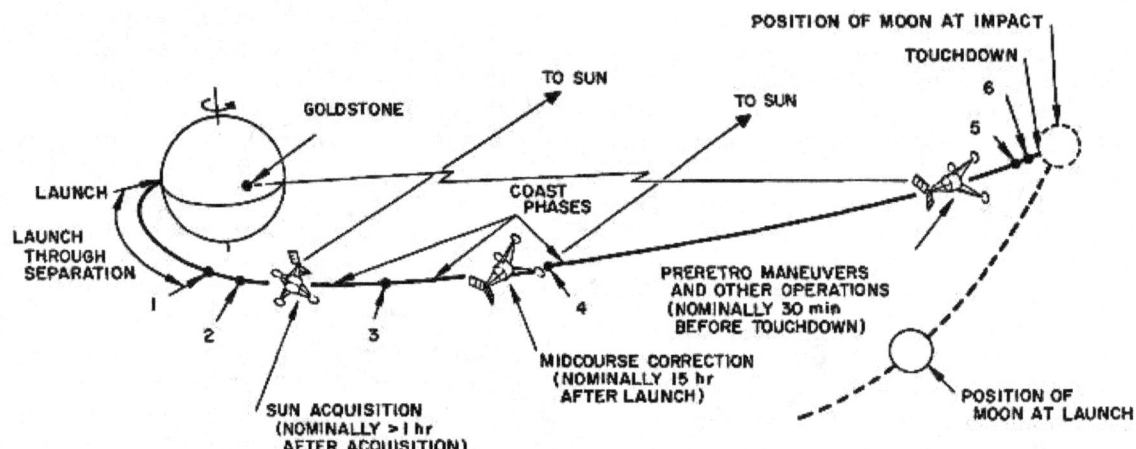

I INJECTION AND SEPARATION
2 INITIAL DSIF ACQUISITION
3 STAR ACQUISITION AND VERIFICATION
   (NOMINALLY 15 hr AFTER LAUNCH)
4 REACQUISITION OF SUN AND STAR
   (IMMEDIATELY AFTER MIDCOURSE CORRECTION)
5 RETRO PHASE (INITIATED AT NOMINAL RANGE
   OF 60 mi FROM MOON)
6 VERNIER DESCENT (NOMINALLY LAST 35,000 ft
   OF FLIGHT)

Fig. VII-2. Earth–moon trajectory and nominal events

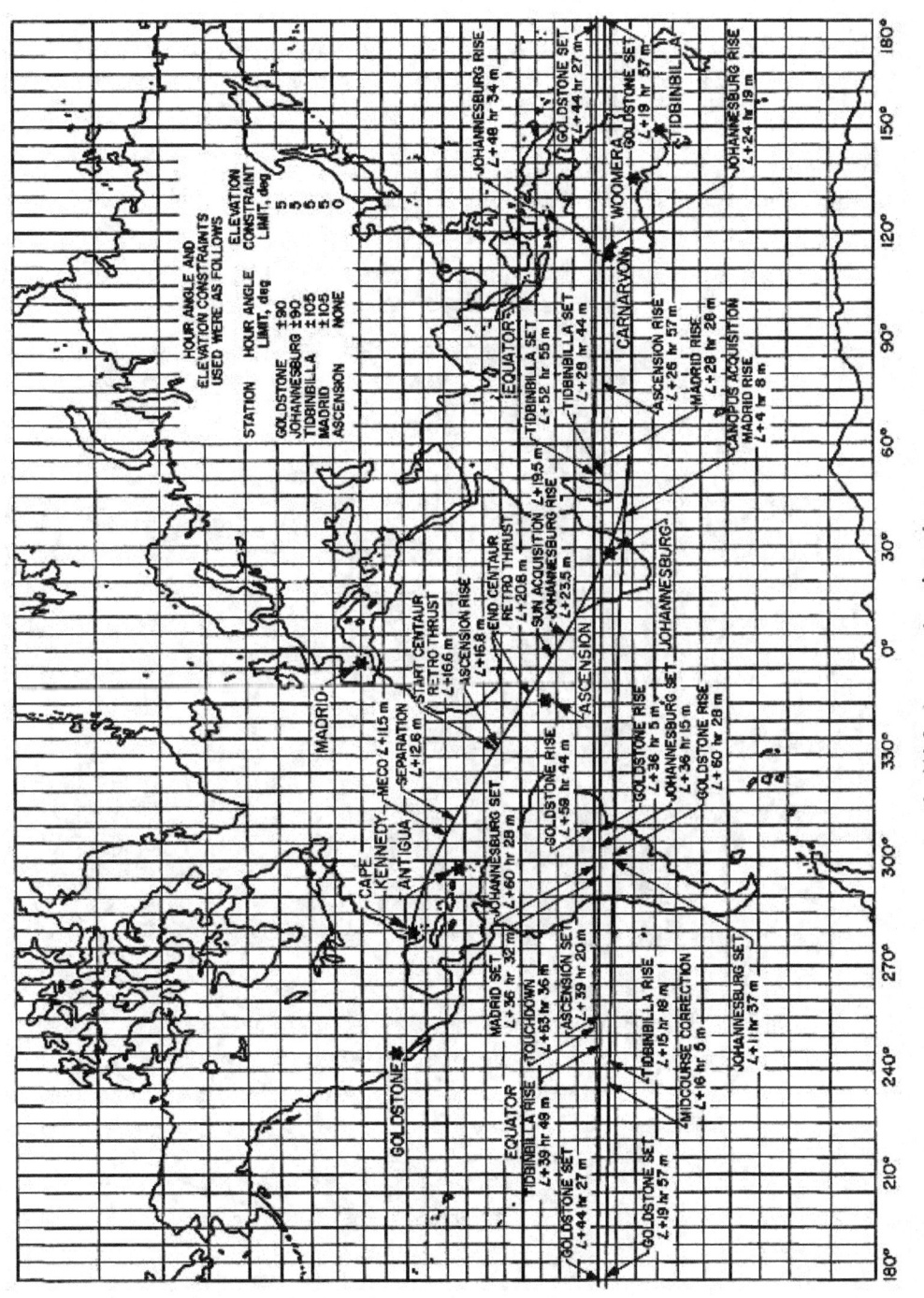

Fig. VII-3. Surveyor I earth track

longitude) is shown in Fig. VII-4. The uncorrected, unbraked impact point is located on the western edge of Oceanus Procellarum, west of the crater Hansteen. The original aim point is approximately 400 km to the northeast, just north of the crater Flamsteed.

Figure VII-5 illustrates the *Centaur* and *Surveyor* trajectories. The projection of each trajectory is plotted on the earth's equatorial plane. The best estimate of the *Centaur* injection conditions was obtained from ETR. These conditions were computed in-flight based upon *Centaur* post-retro data. A mission design constraint states that the *Centaur/Surveyor* separation distance must be at least 336 km by 5 hr after injection to eliminate possible *Centaur* interference during Canopus acquisition. The required separation distance was reached

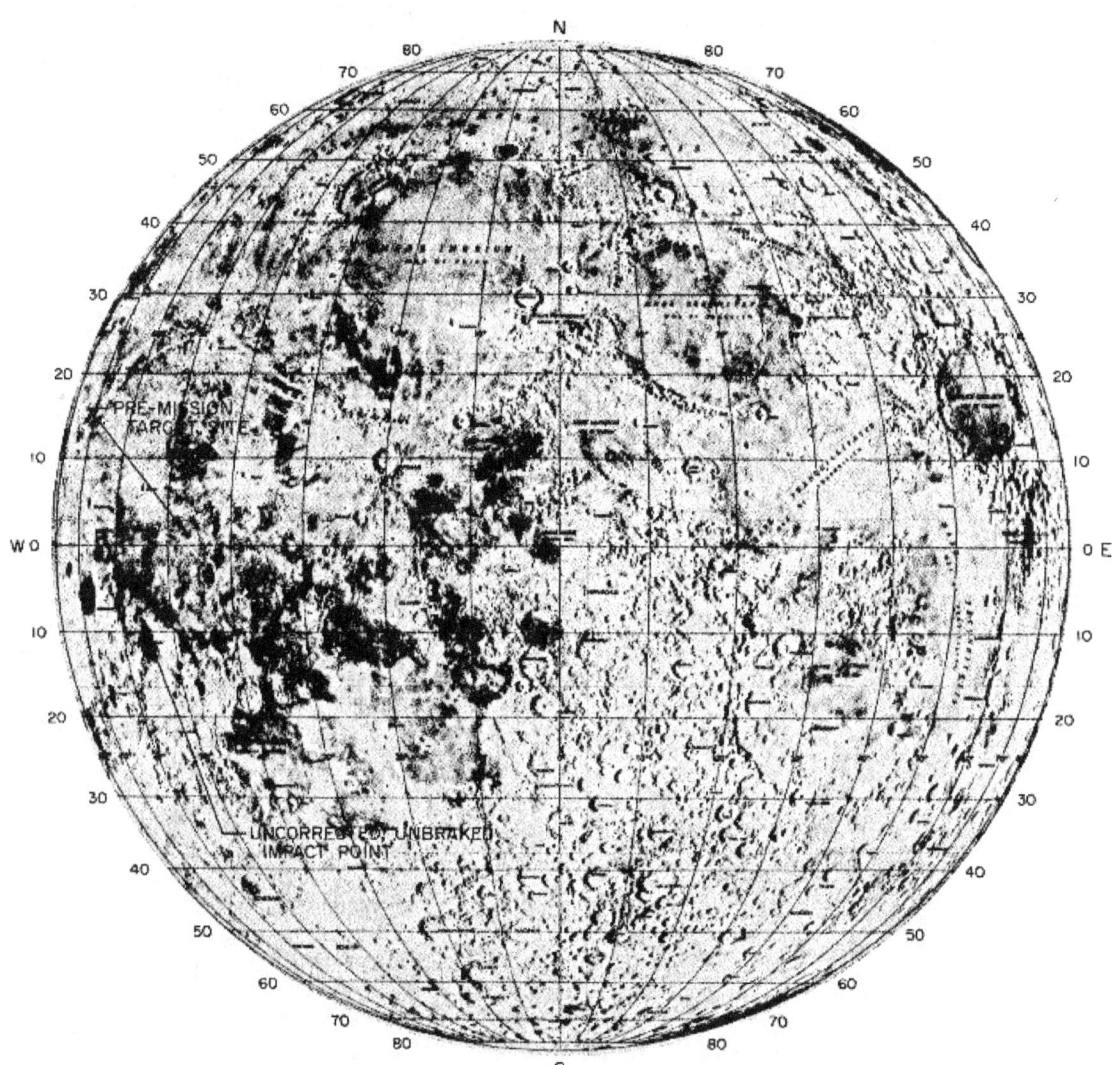

Fig. VII-4. *Surveyor I program target and uncorrected impact points*

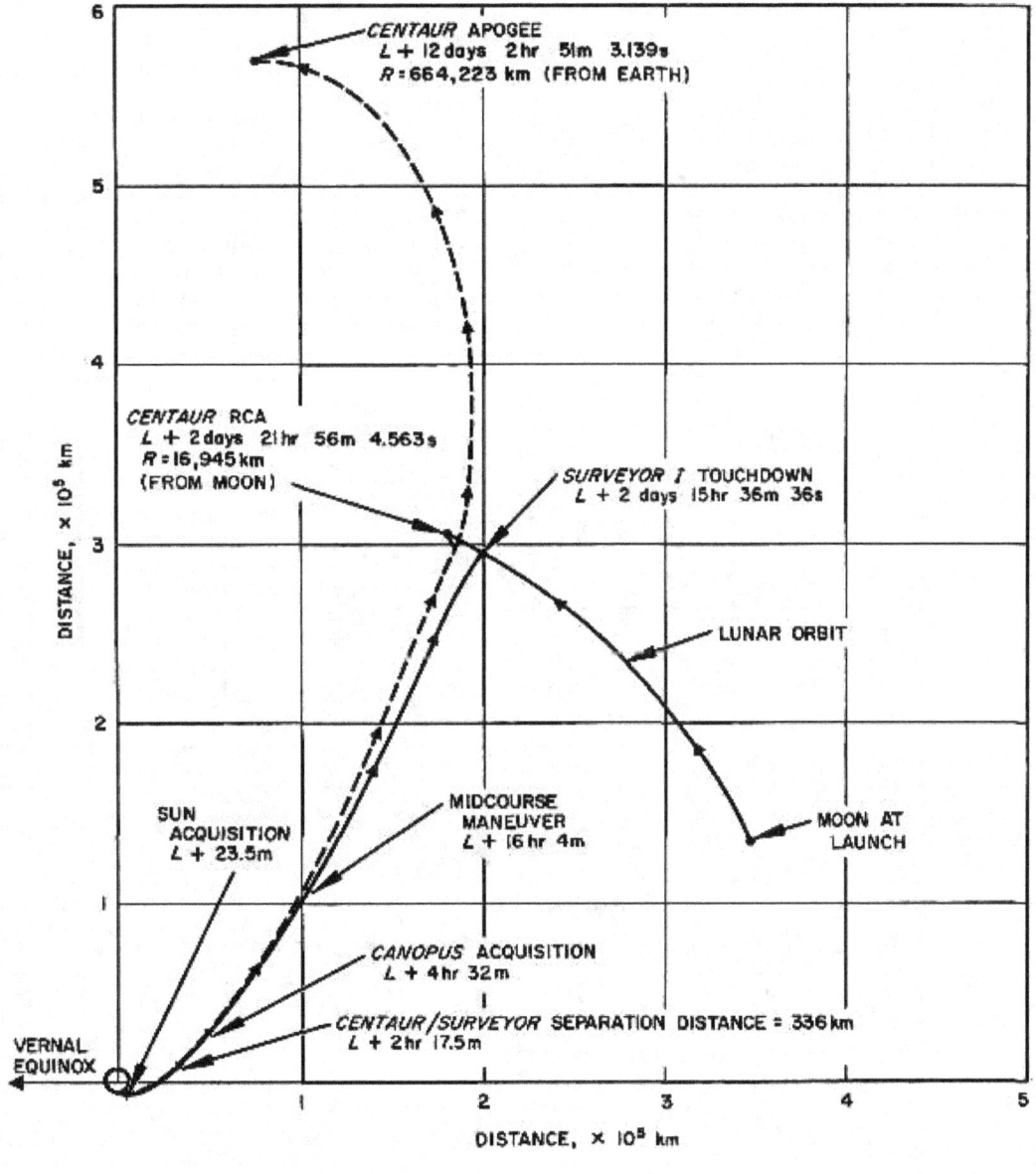

Fig. VII-5. *Surveyor* and *Centaur* trajectories in earth's equatorial plane

2 hr and 17.5 min after launch. The *Centaur* passed above and behind the moon about 6 hr and 20 min after *Surveyor I* touchdown.

C. Midcourse Maneuver Phase

The original aim point was selected assuming the 99% landing dispersion to be a 50-km-radius circle on the lunar surface. However, primarily because of the small midcourse correction required, the 99% dispersion computed in-flight was an ellipse having a semimajor axis of 38.7 km and a semiminor axis of 28.7 km. Because of this smaller dispersion, it was decided to bias the aiming point to the north approximately 0.92 deg to minimize the chance of landing in the craters Flamsteed or Flamsteed E. Figure VII-6 shows the initial aim point, the final aim point (−2.33 deg latitude, 316.17 deg longitude), the 99% dispersion associated with premidcourse tracking and execution errors, and the actual soft-landing

site. The latitude and longitude for these locations as well as for the uncorrected impact point are given below:

	Latitude, deg	Longitude, deg
Original aim point	− 3.25	316.17
Uncorrected impact point	−11.425	305.853
Final aim point	− 2.33	316.17
Actual landing site (best estimates)		
Based on flight tracking data	− 2.41	316.65
Based on post-touchdown tracking data	− 2.58	316.65

The midcourse correction, computed to enable the spacecraft to soft-land at the new aiming point, was 20.35 m/sec. This correction was executed upon ground command at approximately 06:45 GMT on May 31, 1966. The resulting soft-landing site is estimated to be −2.58 deg latitude and 316.65 deg longitude, well within the

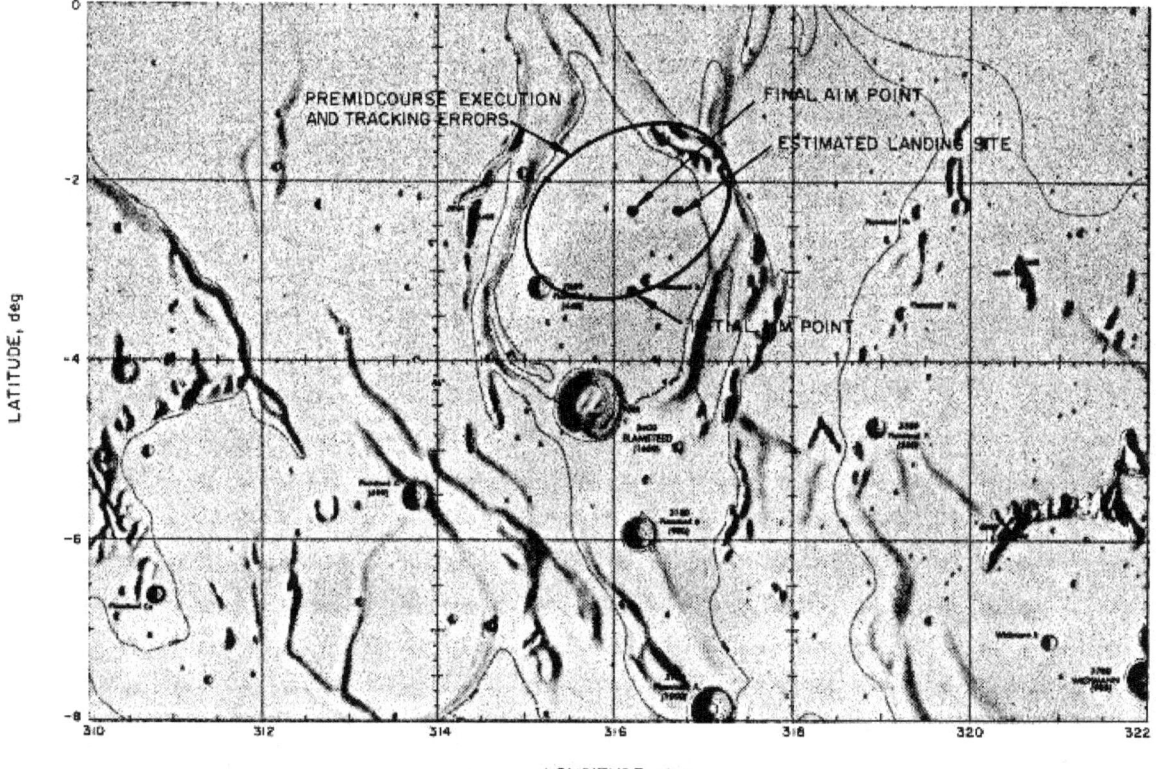

Fig. VII-6. Surveyor I landing location

3-σ dispersions predicted prior to the correction. Dispersions on the actual landing site are presently estimated to be 5 km in latitude and 2 km in longitude.

The maximum midcourse correction capability, as a function of the unbraked impact speed, is shown in Fig. VII-7. The expected 3-σ *Centaur* injection guidance dispersions and the effective lunar radius are also shown. The midcourse capability contours are in the conventional R-S-T coordinate system.*

*Kizner, W. A., *A Method of Describing Miss Distances for Lunar and Interplanetary Trajectories*, External Publication 674, Jet Propulsion Laboratory, Pasadena, August 1, 1959.

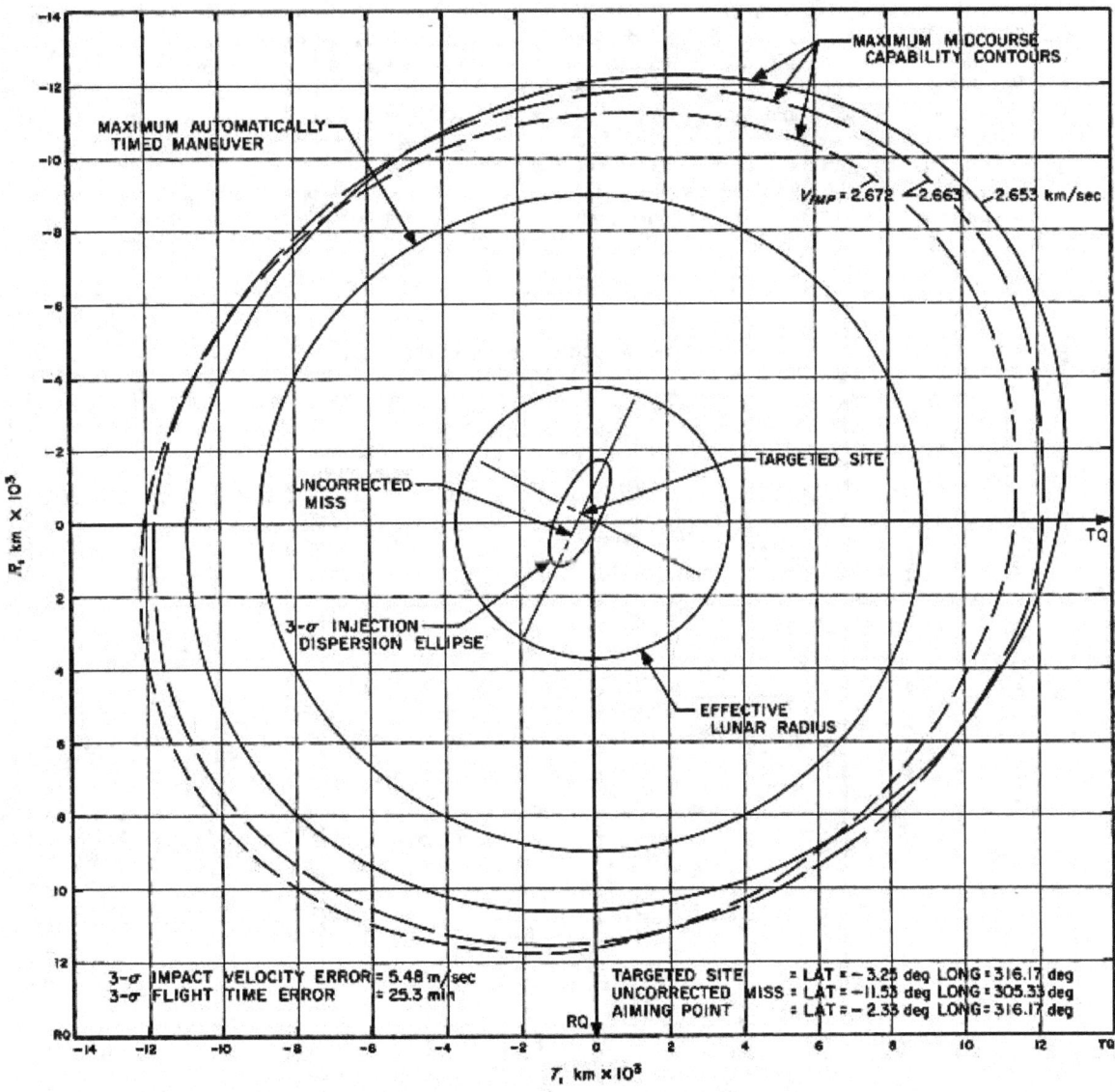

Fig. VII-7. Midcourse capability contours for May 30 launch

The maneuver execution time of 15.85 hr after injection was chosen. This time allowed 4 hr 36 min of premidcourse and 3 hr 49 min of postmidcourse visibility from the Goldstone DSS 11 tracking facility.

The midcourse velocity component required to correct the "miss only" was 3.74 m/sec, computed to correct to the final aim point. This component is referred to as the critical component. The velocity component normal to the critical component is referred to as the noncritical component since it does not affect the miss to first order. Figure VII-8 presents the variations in flight time, main retro burnout velocity, and vernier propellant margin with the noncritical velocity component U_3. The propellant margin and flight time were acceptable over the wide range of values within the limits shown. Flight control stability considerations, however, made main retro burn-

out velocity below 450 ft/sec highly desirable. A positive increment of 20 m/sec in the noncritical direction was selected as a good compromise, resulting in a nominal burnout velocity of 392 ft/sec.

If the maneuver strategy had been simply to correct miss and flight time to the new aim point, the required noncritical component would have been 4.3 m/sec, giving a total correction of 6.1 m/sec. However, to properly evaluate the performance of the *Centaur* guidance system, the original aiming point must be used in computing the correction, in which case the "miss only" correction is 3.6 m/sec and the miss-plus-flight-time correction is 5.9 m/sec.

Table VII-1 presents the premidcourse and postmidcourse injection and terminal conditions.

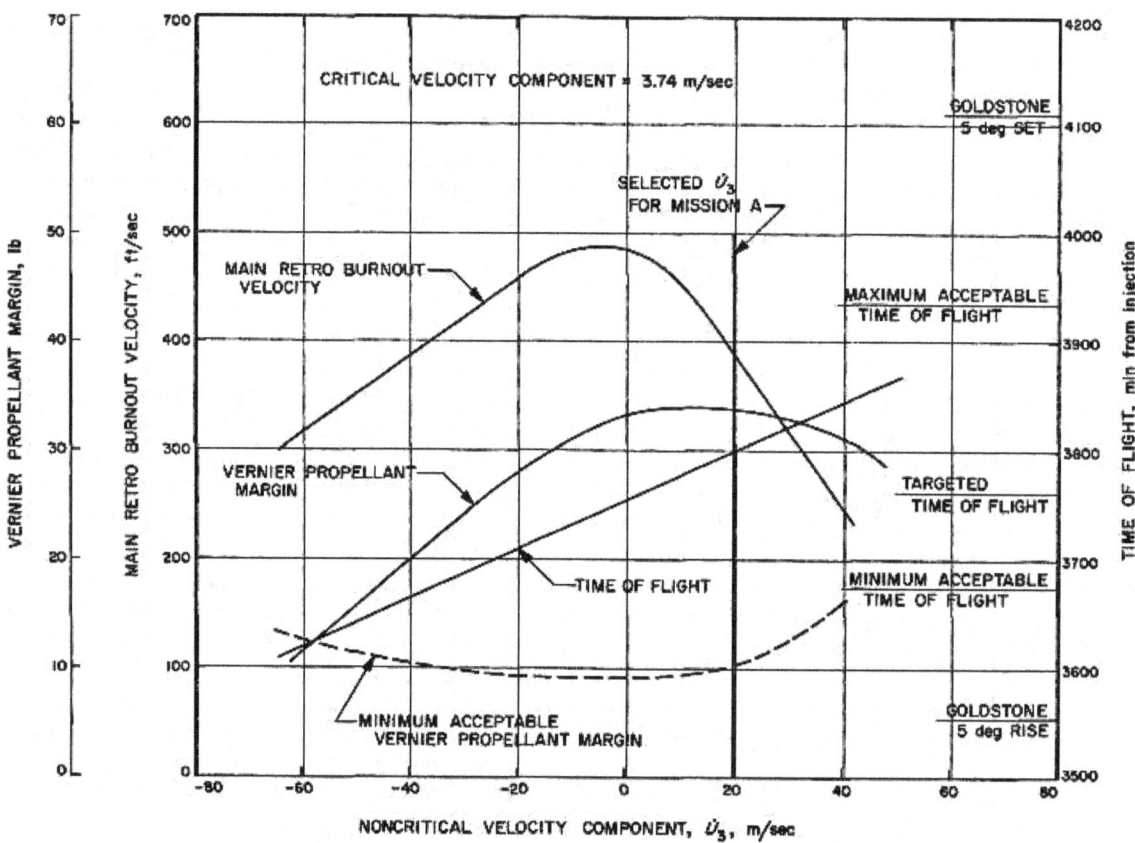

Fig. VII-8. Effect of noncritical velocity increment on terminal descent parameters

Table VII-1. Premidcourse and postmidcourse injection and terminal conditions

Premidcourse injection conditions, May 30, 1966, 1e:53:57.385 GMT

Coordinate* system						
Inertial Cartesian	X = 2813.6136 km	Y = 5574.0461 km	Z = 1979.7524 km	DX = −8.3688738 km/sec	DY = 5.5320036 km/sec	DZ = −4.4497759 km/sec
Inertial spherical	RAD = 6550.2540 km	DEC = 17.592224 deg	RA = 63.216706 deg	VI = 10.974589 km/sec	PTI = −1.2120495 deg	AZI = 114.75517 deg
Earth-fixed spherical	RAD = 6550.2540 km	LAT = 17.592224 deg	LON = 342.14293 deg	VE = 10.562934 km/sec	PTE = −1.2592931 deg	AZE = 115.78965 deg
Orbital elements	C3 = −1.2639824 km²/sec²	ECC = 0.97923824	INC = 30.044723 deg	TA = −2.4498112 deg	LAN = −276.45973 deg	APF = 145.31695 deg

Premidcourse encounter conditions, June 2, 1966, 05:28:59.985 GMT

Coordinate system						
Selenocentric	RAD = 1738.09 km	LAT = −1.419473 deg	LON = 305.85900 deg	VP = 2.6649810 km/sec	PTP = −79.992989 deg	AZP = 264.77994 deg
Miss parameter, earth equator	BTQ = −585.25804 km	BRQ = 305.93221 km	B = 660.39495 km			
Miss parameter, moon equator	BTT = −656.92308 km	BRT = 67.634913 km	B = 660.39365 km			

Postmidcourse injection conditions, May 31, 1966, 06:45:53.800 GMT

Coordinate system						
Inertial Cartesian	X = −103401.67 km	Y = −107006.61 km	Z = −66923.279 km	DX = −0.88951855 km/sec	DY = −1.5378124 km/sec	DZ = −0.62099156 km/sec
Inertial spherical	RAD = 1629.1439 km	DEC = −24.023111 deg	RA = 225.98155 deg	VI = 1.8819512 km/sec	PTI = 76.179004 deg	AZI = 72.573537 deg
Earth-fixed spherical	RAD = 1629.1439 km	LAT = −24.023111 deg	LON = 236.18758 deg	VE = 10.581803 km/sec	PTE = 9.9447604 deg	AZE = 270.74016 deg
Orbital elements	C3 = −1.3516429 km³/sec³	ECC = 0.976691524	INC = 29.371361 deg	TA = 159.89571 deg	LAN = 278.34842 deg	APF = −144.00209 deg

Postmidcourse encounter conditions, June 2, 1966, 06:15:14.632 GMT

Coordinate system						
Selenocentric	RAD = 1735.6000 km	LAT = −2.3555486 deg	LON = 316.54097 deg	VP = 2.6553638 km/sec	PTP = −83.868529 deg	AZP = 343.30690 deg
Miss parameter, earth equator	BTQ = −253.86534 km	BRQ = −327.05272 km	B = 414.01824 km			
Miss parameter, moon equator	BTT = −114.52272 km	BRT = −397.86410 km	B = 414.01824 km			

*Holdridge, D. B. Space Trajectories Program for the IBM 7090 Computer, Technical Report No. 32-223, Jet Propulsion Laboratory, Pasadena, March 2, 1962.

D. Terminal Phase

Because of the apparent failure of Omniantenna A to deploy, attitude maneuvers in the terminal phase were selected to optimize signal strength from Omniantenna B. Study of this problem showed that there were two possible maneuver sequences that would maintain relatively high signal strength during and following the maneuvers. The first maneuver sequence was a roll-yaw-roll combination that gave a final spacecraft roll orientation such that the DSIF station was in the most favorable location. The second maneuver sequence was a roll-pitch combination. This maneuver sequence also resulted in an equally favorable final roll orientation. However, the latter maneuver sequence had disadvantages that the first sequence did not have. First, the pitch maneuver channel had not been exercised at midcourse as had the yaw channel. Second, the roll-pitch combination could not be computed directly with the midcourse maneuver analysis computer programs, and thus there could be no compensation for a sensor group deflection of 0.34 deg and a known Y-gyro drift of approximately 0.75 deg/hr. The net result of these two uncompensated error sources was an estimated 0.4-deg offset in the retro thrust vector. With this expected thrust offset, the flight path angle 3-σ uncertainty at the start of the vernier phase was 28 deg as compared to an uncertainty of 20 deg when these errors are compensated.

Because of these disadvantages, the roll-yaw-roll sequence was finally chosen. To offset the increased operational time disadvantage inherent in a three-maneuver sequence, the first maneuver was executed 38 min prior to retro ignition rather than the standard 33 min generally used to minimize gyro drift error. This execution allowed 3 min to complete the first maneuver and 2 min to set up for the second maneuver so that it could be executed at 33 min. The additional 5 min operational time was allowable because, during the first roll maneuver, sun lock was maintained. Thus, pitch and yaw attitude errors owing to gyro drifts were held to zero until the second maneuver was executed. In addition, of course, the gyro drift rates were being compensated for to the extent that they were constant and accurate.

Final attitude maneuver magnitudes were based on measured gyro drift rates of zero in the pitch axis, 0.75 deg/hr in the yaw axis, and 0.2 deg/hr in the roll axis.

Figure VII-9 presents the terminal descent sequence. Table VII-2 tabulates predicted terminal descent parameters. Actual terminal parameters are still being determined.

Table VII-2 of Appendix A gives terminal event times. Actual touchdown time occurred within about 2 sec of the time that had been predicted prior to initiation of the terminal descent sequence.

Table VII-2. Predicted values* of terminal parameters

Terminal parameters	Values
Ignition altitude, ft	246,636
Ignition velocity, ft/sec	8565.3
Main retro burnout altitude, ft	28,571
Main retro burnout velocity, ft/sec	392
Flight path angle, deg	6.02
Nominal propellant consumption (vernier phase), lb	78.2
Propellant margin (touchdown), lb	33.9
Usable touchdown propellant margin required for 99% burnout dispersions, lb	10.5
Landing site	
Latitude, deg	−2.41
Longitude, deg	316.65

*Based on the final postmidcourse computer run of the Terminal Guidance Program, completed approximately 2.5 hr prior to main retro ignition.

E. Landing Site

The estimate of the actual landing site derived using flight tracking data (−2.41 deg latitude and 316.65 deg longitude) is based upon an unbraked impact point of −2.356 deg latitude and 316.642 deg longitude. The terminal guidance program predicted that the difference between unbraked impact and landing would be 1.13 km along the surface in the trajectory plane (i.e., at an azimuth of 343.4 deg). Since landing, a study of terminal descent telemetry data shows that there was an additional 42 ft/sec lateral velocity not considered in the initial terminal guidance solution. This lateral component added another 0.59 km, bringing the total difference to 1.72 km, which is the value used to determine the estimate given above.

Post-landing doppler tracking data is currently being analyzed to locate the landing site more accurately. The present best estimate of the actual landing site, based upon preliminary analysis of post-touchdown tracking data, is −2.58 deg latitude and 316.65 deg longitude. This estimate is in good agreement with landing site determinations based upon photographic data. The use of photographic data to determine landing site location is discussed in Part II of this report.

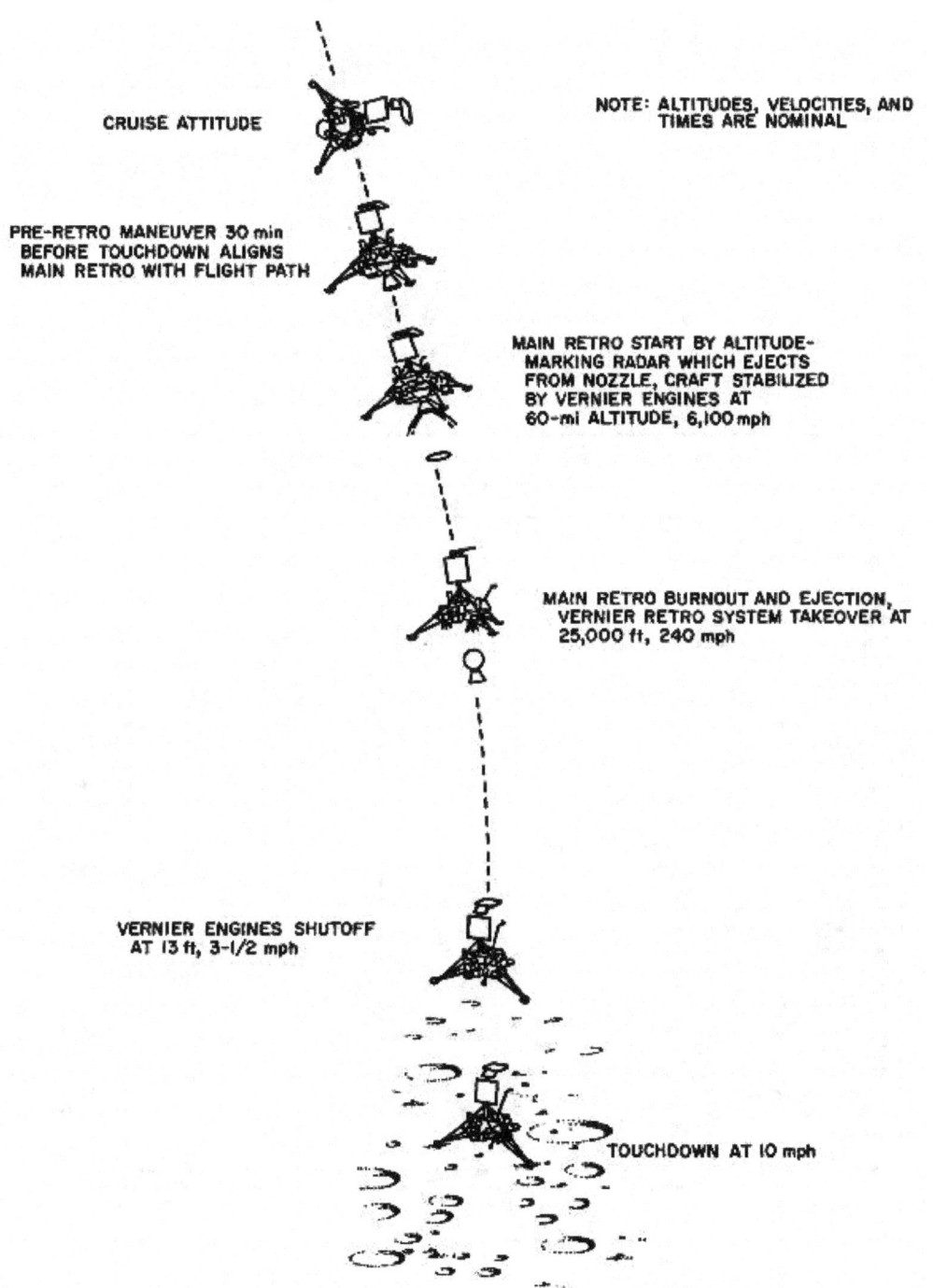

CRUISE ATTITUDE

NOTE: ALTITUDES, VELOCITIES, AND
TIMES ARE NOMINAL

PRE-RETRO MANEUVER 30 min
BEFORE TOUCHDOWN ALIGNS
MAIN RETRO WITH FLIGHT PATH

MAIN RETRO START BY ALTITUDE-
MARKING RADAR WHICH EJECTS
FROM NOZZLE, CRAFT STABILIZED
BY VERNIER ENGINES AT
60-mi ALTITUDE, 6,100 mph

MAIN RETRO BURNOUT AND EJECTION,
VERNIER RETRO SYSTEM TAKEOVER AT
25,000 ft, 240 mph

VERNIER ENGINES SHUTOFF
AT 13 ft, 3-1/2 mph

TOUCHDOWN AT 10 mph

Fig. VII-9. Terminal descent nominal events

APPENDIX A

Surveyor Mission A Events

Table A-1. Mission flight events

Event	Mark No.	Mission time (predicted)[a]	Mission time (actual)	GMT (actual)
Launch to DSIF acquisition				
Launch (2-in. rise)		L+00:00:00.00	L+00:00:00.00	14:41:00.990
Initiate roll program		L+00:00:02		
Terminate roll, initiate pitch program		L+00:00:15		
Mach 1			L+00:00:58	14:41:59
Max. aerodynamic loading			L+00:01:17	14:42:18
Booster engine cutoff (BECO)	1	L+00:02:22.41	L+00:02:22.04	14:43:23.03
Jettison booster package	2	L+00:02:25.51	L+00:02:25.12	14:43:26.11
Admit guidance steering		L+00:02:30		
Jettison Centaur insulation panels	3	L+00:02:56.41	L+00:02:55.84	14:43:56.83
Jettison nose fairing	4	L+00:03:23.41	L+00:03:22.76	14:44:23.75
Start Centaur boost pumps		L+00:03:24.41	L+00:03:23.72	14:44:24.71
Sustainer engine cutoff (SECO)	5	L+00:04:00.67	L+00:03:59.38	14:45:00.37
Atlas/Centaur separation	6	L+00:04:02.50	L+00:04:01.31	14:45:02.30
Centaur main engine ignition (MEIG)	7	L+00:04:12.17	L+00:04:10.86	14:45:11.85
Centaur main engine cutoff (MECO)	8	L+00:11:27.33	L+00:11:29.21	14:52:30.20
Vehicle destruct systems safed by ground command			L+00:11:40.1	14:52:41.1
Extend Surveyor landing legs command	9		L+00:11:55	14:52:56
Extend Surveyor antennas command	10		L+00:12:05	14:53:06
Surveyor transmitter high power on command	11		L+00:12:26	14:53:27
Surveyor/Centaur electrical disconnect	12		L+00:12:31	14:53:32
Surveyor/Centaur separation	13	L+00:12:38.17	L+00:12:36.93	14:53:37.92
Start Centaur 180 deg. turn	14		L+00:12:42	14:53:43
Start Centaur lateral thrust			L+00:13:22	14:54:23
Cutoff Centaur lateral thrust			L+00:13:42	14:54:43
Start Centaur retro (blowdown tanks)	17		L+00:16:36	14:57:37
Sun acquisition cell illuminated			L+00:16:49	14:57:50
Solar panel locked in transit position			L+00:18:19	14:59:20
Sun lock-on completed			L+00:19:33	15:00:34
Cutoff Centaur retro and pwr off	18, 19		L+00:20:47	
Roll axis locked in transit position			L+00:22:19	15:03:20
Initial DSIF acquisition completed (two-way lock)			L+00:27:30	15:08:31

[a] Predicted values from Centaur Monthly Configuration, Performance, and Weight Status Report, Report GDC63-0495-37, Appendix A, Trajectory and Performance Data, General Dynamics/Convair, San Diego. The predicted values were computed postflight utilizing actual launch azimuth, tanked propellant weights and atmospheric data which depend on day and time of liftoff.

Table A-1 (Cont'd)

Event	Mission time (actual)	GMT (actual)
DSIF acquisition to star acquisition		
Initial spacecraft operations		
1. Command transmitter from high to low power	L+00:39:58	15:20:59
2. Command off basic bus accelerometer amplifiers and solar panel deployment logic		
3. Command rock solar panel back and forth to seat locking pin		
4. Command rock roll axis back and forth to seat locking pin	L+00:46:56	15:27:55
5. Perform interrogation of all engineering signal processor data modes at 1100 bit/sec	L+00:53:14 to L+01:46:56	15:34:15 to 15:56:04
Attempt to extend Omnidirectional antenna A by ground command	L+01:40:20 to L+01:45:07	16:21:21 to 16:26:08
Star verification/acquisition		
1. 1100 bit/sec engineering interrogation of modes 4, 2, and 1	L+03:50:53	18:31:54
2. Command transmitter high power turn on prior to star verification (high voltage on at 18:44:24 GMT)	L+04:01:20	18:42:21
3. Command on coast mode commutator	L+04:08:33	18:49:34
4. Command execution of positive roll (roll begun at 18:53:39 GMT)	L+04:09:43	18:50:44
5. Command manual star lock achieved	L+04:32:19	19:13:20
6. Command return to low-power operation	L+04:40:03	19:21:04
Premidcourse coast phase		
Command gyro drift check	L+06:13:01 to L+09:10:09	20:54:02 to 23:51:10
Command bit rate reduction from 1100 to 550 bit/sec	L+10:57:18	01:38:19 (5/31/66)
Low-power engineering interrogation of modes 4 and 2	L+11:59:00 to L+12:06:08	02:40:01 to 02:47:09
Engineering interrogation at 1100 bit/sec with low power	L+13:33:04 to L+13:37:02	04:14:05 to 04:20:03
Command gyro speed check and return to 550 bit/sec telemetry data	L+13:45:03 to L+13:49:16	04:26:04 to 04:30:17
Command return to coast mode commutator data	L+13:52:36 to L+13:52:43	04:33:37 to 04:33:44
Command manual lockon to Canopus to null out star error signal prior to midcourse execution	L+14:14:35	04:55:36
Midcourse correction		
Midcourse correction sequence		
1. Engineering interrogation of modes 4, 2, and 1 at 1100 bit/sec with transmitter low power	L+15:20:14 to L+15:25:45	06:01:15 to 06:06:46
2. Command on high-power transmitter (high voltage on at 06:08:25 GMT)	L+15:25:45	06:06:46
3. Command increase telemetry rate from 550 to 4400 bit/sec	L+15:29:22	06:10:23
4. Command pressurize vernier system (helium)	L+15:38:07	06:19:08
5. Command roll maneuver magnitude and direction (minus roll of 86.5 deg)	L+15:39:22	06:20:23
6. Command roll execution	L+15:49:12	06:30:13
7. Command yaw maneuver magnitude and direction (minus yaw of 57.99 deg)	L+15:53:13	06:34:14
8. Command yaw execution	L+15:53:47	06:34:48
9. Command propulsion strain gauge power on	L+15:57:03	06:38:05
10. Command unlock Engine 1 to permit roll control	L+15:58:06	06:39:07
11. Command thrust phase power on	L+16:02:04	06:43:05

Table A-1 (Cont'd)

Event	Mission time (actual)	GMT (actual)
Midcourse correction (Cont'd)		
12. Command desired thrust duration (20.8 sec)	L+16:02:41	06:43:42
13. Command midcourse thrust execution	L+16:04:02	06:45:03
14. Turn off thrust phase power	L+16:02:25	06:45:26
15. Command off propulsion strain gauge power	L+16:05:21	06:46:22
16. Command operations to obtain coast mode data at 550 bit/sec (for obtaining receiver A AGC data in an attempt to verify condition of Omnidirectional antenna A)	L+16:08:01 to L+16:09:36	06:49:02 to 06:50:37
17. Command reverse yaw maneuver magnitude and direction (plus 57.99 deg)	L+16:10:45	06:51:46
18. Command yaw execution (sun reacquired at 06:54:51 GMT)	L+16:11:54	06:52:55
19. Command reverse roll maneuver magnitude and direction (plus 85.5 deg)	L+16:16:03	06:57:04
20. Command roll execution (Canopus reacquired at approximately 07:00:55 GMT)	L+16:47:01	06:58:02
21. Postmidcourse engineering interrogation of modes 1, 2, and 4	L+16:23:29 to L+16:30:52	07:04:30 to 07:11:53
22. Command return to transmitter low-power	L+16:31:42	07:12:43
23. Command off auxiliary accelerometer amplifier and touchdown strain gauge power	L+16:45:38	07:26:39
Postmidcourse coast phase		
Low-power interrogation of modes 4 and 2 to obtain thermal data	L+17:51:01	08:32:02
Command gyro drift check	L+18:49:13 to L+21:49:15	09:30:14 to 12:30:16
Command start of drift check of roll axis only	L+22:55:16	13:36:17
Low-power interrogation of modes 4 and 2 to obtain thermal data	L+25:32:27 to L+25:46:20	16:13:28 to 16:27:41
Low-power interrogation of modes 4 and 2 to obtain thermal data	L+29:51:41 to L+30:02:43	20:32:42 to 20:43:44
Command termination of roll drift check to null out accumulated error produced by drift	L+32:52:37	23:37:38
Low-power interrogation	L+37:24:54 to L+37:30:53	04:05:55 to 04:11:54 (6/1/66)
Command gyro drift check for pitch and yaw gyros; roll gyro drift check allowed to continue	L+37:37:38 to L+39:34:56	04:18:39 to 06:25:57
Attempt to extend Omnidirectional antenna A by ground command	L+39:50:16	06:31:17
Command gyro drift check (roll gyro drift check also terminated)	L+40:53:14 to L+43:11:58	07:34:15 to 09:52:59
Command start of new gyro drift check	L+43:14:41	09:55:42
Low-power interrogation of modes 4 and 2	L+43:17:44 to L+43:26:22	09:58:45 to 10:07:23
Command power mode cycling remaining in auxiliary battery mode to check rate of temperature rise of auxiliary battery	L+45:05:33 to L+45:36:56	11:46:34 to 12:17:57
Command high current mode on—both batteries directly on line	L+45:36:56	12:17:57
Command vernier tanks' thermal control on	L+45:56:57	12:37:58
Command termination of yaw and pitch gyro drift check; roll gyro drift check allowed to continue	L+46:24:02	13:05:03
Command return to auxiliary battery mode to increase rate of temperature rise of auxiliary battery	L+49:06:13	15:47:14

Table A-1 (Cont'd)

Event	Mission time (actual)	GMT (actual)
Postmidcourse coast phase (Cont'd)		
Low-power interrogation of modes 4 and 2	L+49:07:19 to L+49:21:36	15:48:20 to 16:02:37
Command power mode cycling, with return to auxiliary battery (to insure proper auxiliary battery temperature rise)	L+49:48:44 to L+49:53:17	16:29:45 to 16:34:18
Low-power interrogation of modes 4 and 2	L+52:03:15 to L+52:20:19	18:44:16 to 19:01:20
Command termination of roll gyro drift check to null accumulated error	L+52:34:13	19:15:14
Command start of new gyro drift check	L+52:39:42	19:20:43
Command termination of yaw and pitch drift check to null out accumulated error	L+55:00:50	21:41:51
Low-power interrogation of modes 4 and 2	L+56:38:52 to L+56:54:06	23:19:53 to 23:35:07
Command power mode cycling, with return to auxiliary battery mode (to insure proper auxiliary battery temperature rise)	L+57:29:40 to L+57:43:03	00:10:41 to 00:24:04 (6/2/66)
Command on survey TV electronic thermal control power	L+58:35:41	01:16:42
Command both batteries on line in preparation for terminal descent	L+58:41:59	01:23:00
Command termination of roll gyro drift check to null accumulated error	L+59:10:22	01:51:23
Low-power interrogation of mode 4 only	L+59:14:23 to L+59:19:22	01:55:24 to 02:00:23
Low-power interrogation of mode 4 only	L+60:26:59 to L+60:35:53	03:08:00 to 03:16:54
Engineering interrogation of modes 4, 2, and 1 with low power at 550 bit/sec	L+61:14:23 to L+61:20:06	03:55:24 to 04:01:07
Command VCXO check	L+61:24:41 to L+61:28:14	04:05:42 to 04:09:15
Command vidicon temperature control on survey camera		No time confirmation
Terminal descent		
Terminal descent preparation commands 1. Interrogation of modes 4, 2, and 1 and command on high-power transmitter (at 05:20:18 GMT) and command adjustment of telemetry rate to 1100 bit/sec (at 05:21:27); command summing amplifiers off (to turn off presumming amplifier) and command on phase summing amplifier	L+62:37:06 to L+62:44:36	05:18:07 to 05:25:37
2. Command propulsion strain gage power on	L+62:44:36	05:25:37
3. Command touchdown strain gage power on and subcarrier oscillators on	L+62:45:13	05:26:14
4. Command transponder power off (one-way lock achieved)	L+62:48:46 to L+62:51:08	05:29:47 to 05:32:09
5. Command roll maneuver magnitude and direction (plus 89.3 deg)	L+62:51:08	05:32:09

Table A-1 (Cont'd)

Event	Mission time (actual)	GMT (actual)
Thermal descent (Cont'd)		
6. Command roll execution	L+62:55:45	05:36:46
7. Command yaw magnitude and direction (plus 59.9 deg)	L+62:59:43	05:40:44
8. Command yaw execution (retro thrust direction aligned properly at approximately 05:43:47 GMT)	L+63:00:46	05:41:47
9. Command roll maneuver magnitude and direction (plus 94.1 deg)	L+63:03:28	05:44:29
10. Command roll execution	L+63:04:16	05:45:17
11. Command vernier thrust level (200 lb) and retro phase and delay between AMR "mark" and vernier ignition (7.825 sec)	L+63:09:22	05:50:23
12. Command on mode 2 data	L+63:19:33	06:00:34
13. Command retro sequence mode on	L+63:25:26	06:06:27
14. Command vernier lines and tanks thermal control off	L+63:25:51	06:06:52
15. Command AMR power on	L+63:28:56	06:09:57
16. Command thrust phase power on	L+63:29:56	06:10:57
17. AMR enabled by command	L+63:31:56	06:12:57
18. Backup AMR "MARK" time commanded	L+63:33:37	06:14:38
19. AMR "mark" signal	L+63:33:39.7	06:14:40.7[b]
Terminal descent automatic events		
1. Vernier engines ignition	L+63:33:47.6	06:14:48.6[b]
2. Main retro motor ignition	L+63:33:48.6	06:14:49.65[b]
3. Doppler and altimeter radars (RADVS) power on	L+63:33:49.2	06:14:50.2[b]
4. Main retro motor burnout	L+63:34:27.8	06:15:28.85[b]
5. Main retro motor eject	L+63:34:40	06:15:41[b]
6. Main retro motor separation	L+63:34:41	06:15:42[b]
7. Doppler (RADVS) control enabled	L+63:34:43	06:15:44[b]
8. Beam 3 loses lock	L+63:34:43	06:15:44[b]
9. Beam 3 regains lock	L+63:34:45	06:15:46[b]
Command on mode 3 data	L+63:35:00	06:16:01
Command presumming amplifier on (to get touchdown strain gage data)	L+63:35:01	06:16:02
10. 1000 ft mark	L+63:36:15	06:17:16[b]
11. 10 ft/sec mark	L+63:36:28	06:17:29[b]
12. 14 ft mark	L+63:36:34	06:17:35[b]
13. Touchdown (indicated by retro accelerometer)	L+63:36:36	06:17:37[b]

[b] Time telemetry data were received in performance analysis area at Space Flight Operations Facility (i.e., includes transit time from spacecraft to ground, SFOF processing delay, and commutation delay.

Table A-2. Predicted* view period summary

Station	Event	May-June 1966	GMT	Station	Event	May-June 1966	GMT
DSS 51, Johannesburg	5 deg elevation rise	30	15:04:30	DSS 42, Tidbinbilla	5 deg elevation rise	1	06:29:44
DSS 11, Goldstone	5 deg elevation rise	31	02:08:40	DSS 11, Goldstone	5 deg elevation set	1	11:08:04
DSS 51, Johannesburg	90 deg hour angle set	31	02:16:36	DSS 51, Johannesburg	270 deg hour angle rise	1	15:14:32
DSS 42, Tidbinbilla	5 deg elevation rise	31	05:59:23	DSS 42, Tidbinbilla	5 deg elevation set	1	19:36:12
DSS 11, Goldstone	5 deg elevation set	31*	10:33:47	DSS 11, Goldstone	5 deg elevation rise	2	02:25:13
DSS 51, Johannesburg	270 deg hour angle rise	31	15:00:04	DSS 51, Johannesburg	90 deg hour angle set	2	03:09:12
DSS 42, Tidbinbilla	5 deg elevation set	31	19:25:16	DSS 42, Tidbinbilla	5 deg elevation rise**	2	06:38:55
DSS 11, Goldstone	5 deg elevation rise	1	02:23:28	DSS 11, Goldstone	5 deg elevation set**	2	11:25:08
DSS 51, Johannesburg	90 deg hour angle set	1	02:58:45				

*Based upon best set of postinjection trajectory dated.
**View periods of moon's center.

Table A-3. Lunar operations sequences[a]

FIRST LUNAR DAY

June 2, 1966
Post-touchdown standard sequence for turning off power to systems used during terminal descent.
First pictures (200-line) using standard sequence (survey of Pads 2 and 3).
Solar panel and high-gain antenna erection test.
First 600-line pictures (360 deg constant range survey).
Wide-angle survey of one sector (54 deg).
Narrow-angle survey of two sectors (18 deg each).

June 3, 1966
Wide-angle survey of five sectors (54 deg each—360 deg total).
Survey of Pad 2 area.
Narrow-angle surveys of 14 sectors (18 deg each).
Color survey of Pad 2.

June 4, 1966
DSS 51 video training survey (3 pictures).
Surveys of Pads 2 and 3 and crushable block 1 and 2 imprints.
Color survey.
DSS 42 video training survey.
N₂ gas jet/surface reaction study.

June 5, 1966
Surveys of Pad 2 color test chart, Pads 2 and 3, and crushable block 3 imprint.
Narrow-angle survey of 10 sectors (18 deg each).
Color survey of Rock A.
Wide-angle survey of 4 sectors (54 deg each).
Mapping survey of star Sirius.
Survey of Pads 2 and 3.
DSS 42 200-line TV thermal study.

June 6, 1966
(Lunar noon 06:17 GMT)
Surveys of Pad 2 color test chart, Pads 2 and 3, crushable block 1 and 3 imprints.
N₂ gas jet/surface reaction study.
Wide-angle color survey (360 deg total).
Mapping survey of stars Sirius and Canopus.
Wide-angle color survey (360 deg total).
Survey for earth search.

June 7, 1966
Surveys of Pad 2 color test chart, and Pads 2 and 3.
Wide-angle survey (360 deg total).
Narrow-angle color survey of 6 sectors (18 deg each).
Surveys of Pad 2 color test chart, and Pads 2 and 3.

*Other than routine engineering interrogations. All picture sequences are 600-line unless otherwise noted.

Table A-3 (Cont'd)

FIRST LUNAR DAY

June 8 and 9, 1966
No operations during lunar noon.

June 10, 1966
Surveys of Pad 2 color test chart.
Wide-angle survey (360 deg total).
Surveys of Rocks A and B (color), helium bottle (color), focus ranging profile, and crushable block 1 imprint.
Narrow-angle survey of 9 sectors (18 deg each).
Search survey of planet Venus.

June 11, 1966
Survey of Pad 2 color test chart.
Wide-angle survey (360 deg total).
Narrow-angle survey (360 deg total).
Survey of crushable block 1 imprint, Omniantenna B locking strut and auxiliary battery.
Search survey of star Arcturus.
Survey of Compartment A top, and Omniantenna B color test chart.
Narrow-angle color survey of one sector.
Color survey of Compartment B top.
Surveys of Landing Legs 2 and 3.
Mapping survey of star Sirius.

June 12, 1966
Survey of Omniantenna B color test chart.
Wide-angle survey (300 deg total).
Narrow-angle survey (330 deg total).
Narrow-angle color survey of Pads 2 and 3.
Three focus ranging profiles.
Horizon scan, and wide- and narrow-angle azimuth scans.

June 13, 1966
Wide- and narrow-angle horizontal surface scan.
Survey of crushable block 3 imprint.
Narrow-angle surveys of 2 sectors.
Survey of Omniantenna B color test chart.
Wide-angle survey (300 deg total).
Narrow-angle survey (270 deg total).

June 14, 1966
Survey of Omniantenna B color test chart.
Wide-angle survey (360 deg total).
Narrow-angle survey of 5 sectors (18 deg each).
Horizon scans at sunset.
(Start lunar night 15:12 GMT)
Survey of solar corona.
Pad 2 earthshine picture (4-min exposure).

June 15 and 16, 1966
Engineering interrogations only until spacecraft telemetry was commanded off at 20:31 GMT

SECOND LUNAR DAY

June 28 to July 6, 1966
Spacecraft did not respond to commands.

July 7, 1966
Solar panel positioning for increased charge rate.
Camera status verification (24 pictures).

July 8, 1966
Surveys of Pad 2, Compartment A top, and crushable block imprint.
Attempt to fire vernier engines.
Horizon scans at three azimuths.
Surveys of Pad 3, crushable block 3 imprint, and auxiliary battery.
Approach TV camera turn-on (no video transmitted).

July 9 to 11, 1966
No camera operation owing to excessive battery temperature.

July 12, 1966
Survey of rock B.
Horizon scan.
Wide-angle survey (360 deg total).
Repeated surveys of Compartments A and B tops.
Repeated surveys of advancing spacecraft shadow.
Repeated surveys of crushable block 3 imprint.
Surveys of Pads 2 and 3.
Survey of Omniantenna B color test chart.

July 13, 1966
DSS 42 and DSS 51 surveys of spacecraft shadow at hourly intervals.
Surveys of Omniantenna B color test chart, and Pads 2 and 3.
Mirror/filter wheel dust determination.
Narrow-angle survey of 4 sectors.
Survey of Compartment A top.
Narrow-angle survey of 2 sectors.
Survey of Pad 2.

July 14, 1966
DSS 51 solar corona survey attempt (200-line).

APPENDIX B
Surveyor Spacecraft Configuration

POSTLANDING CONFIGURATION

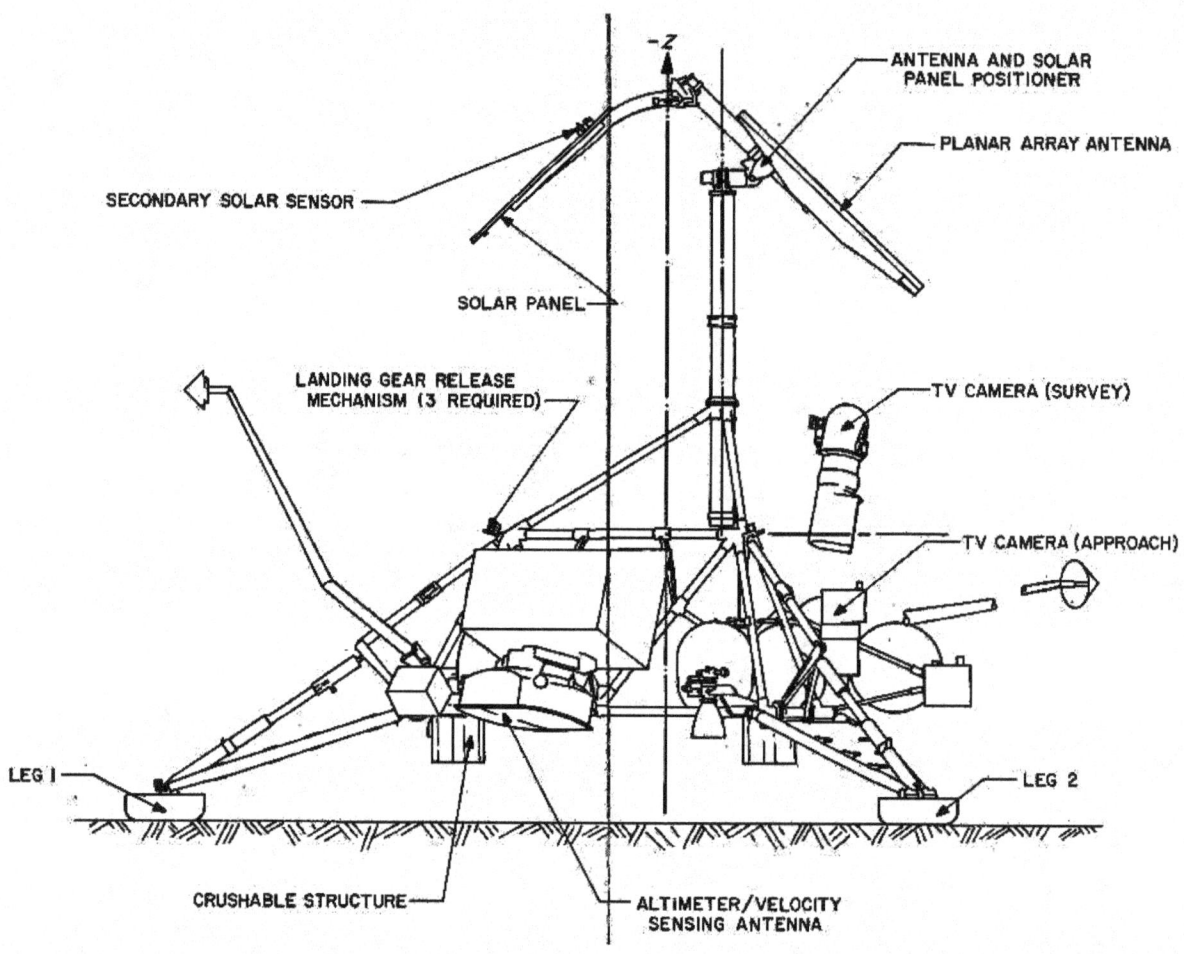

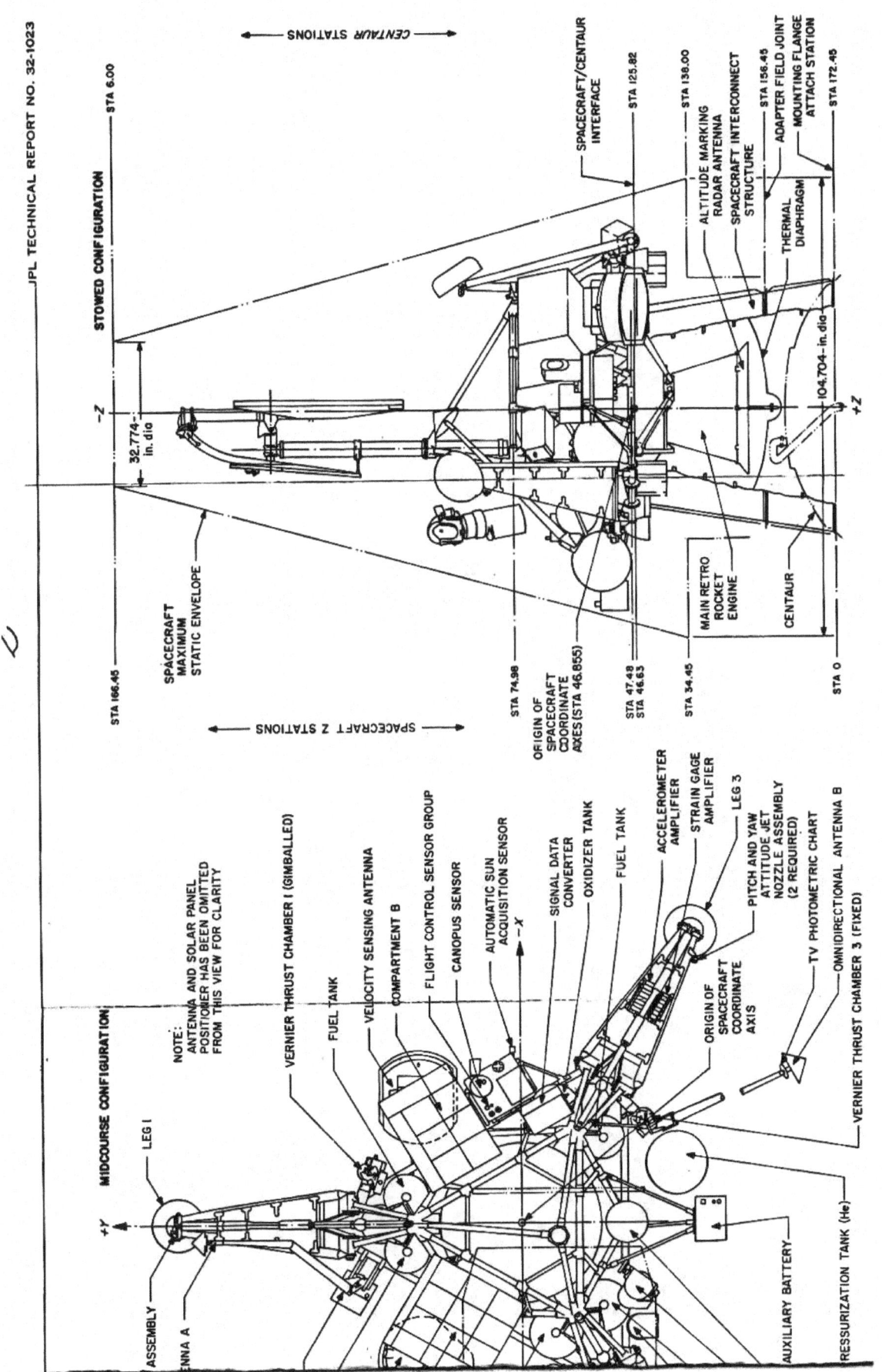

APPENDIX C
Spacecraft Temperature Histories

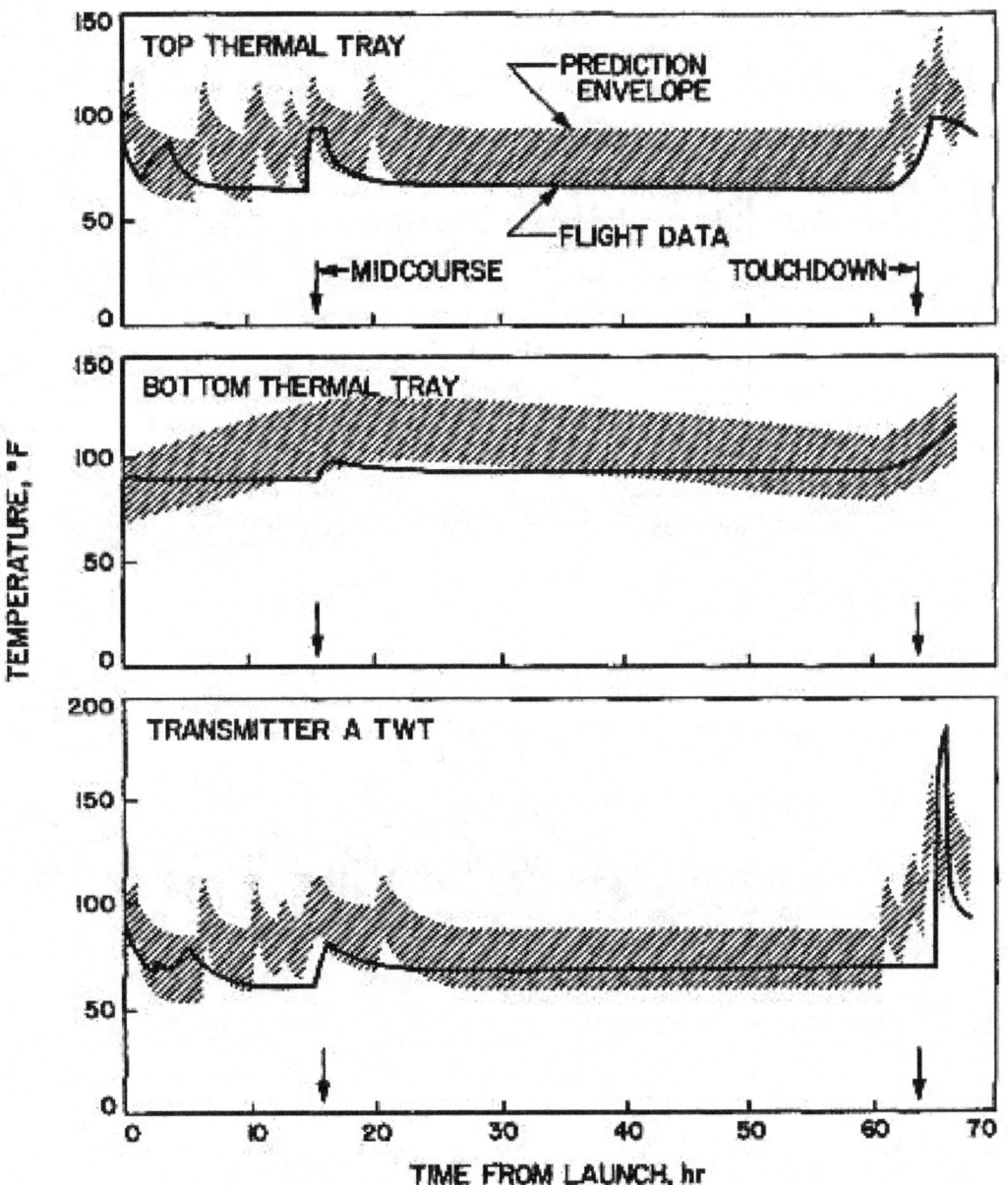

Fig. C-1. Compartment A transit temperatures

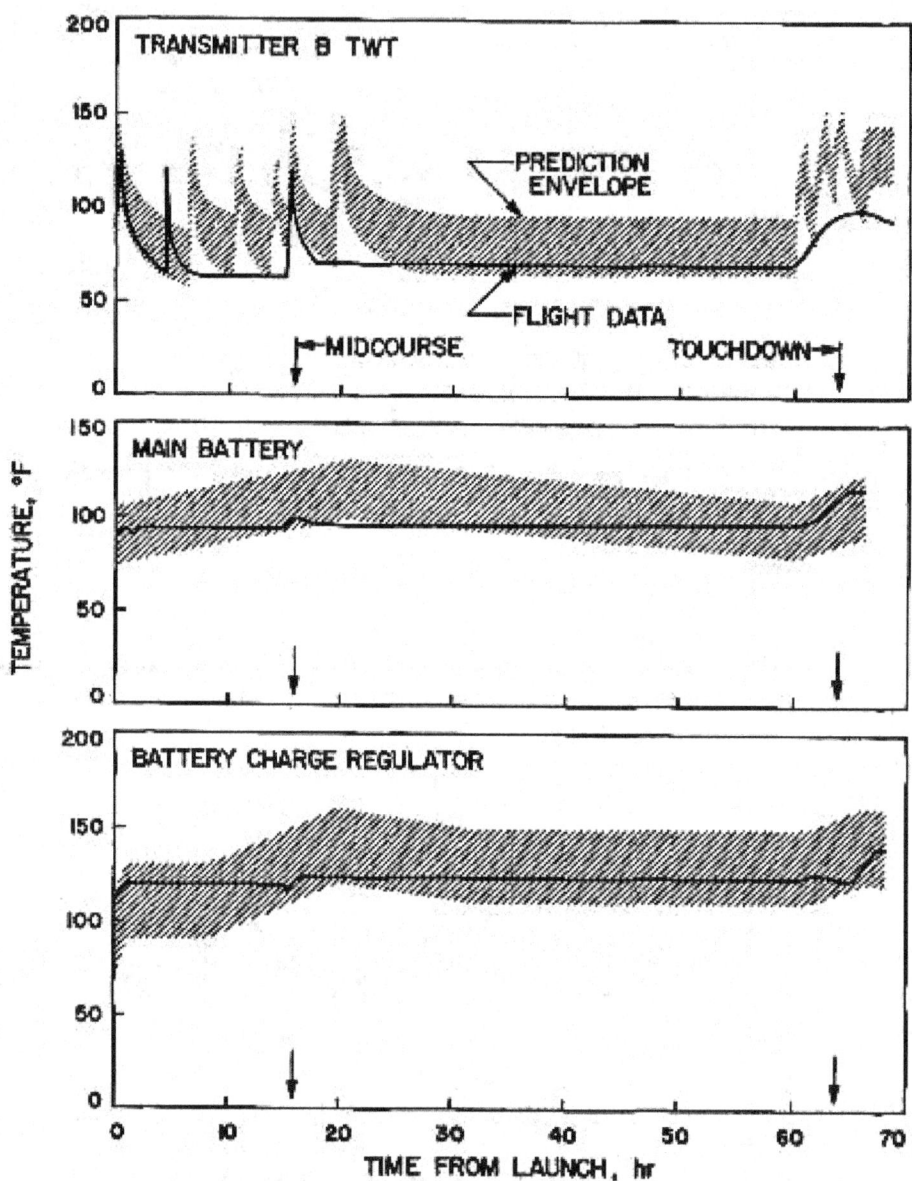

Fig. C-1 (Cont'd)

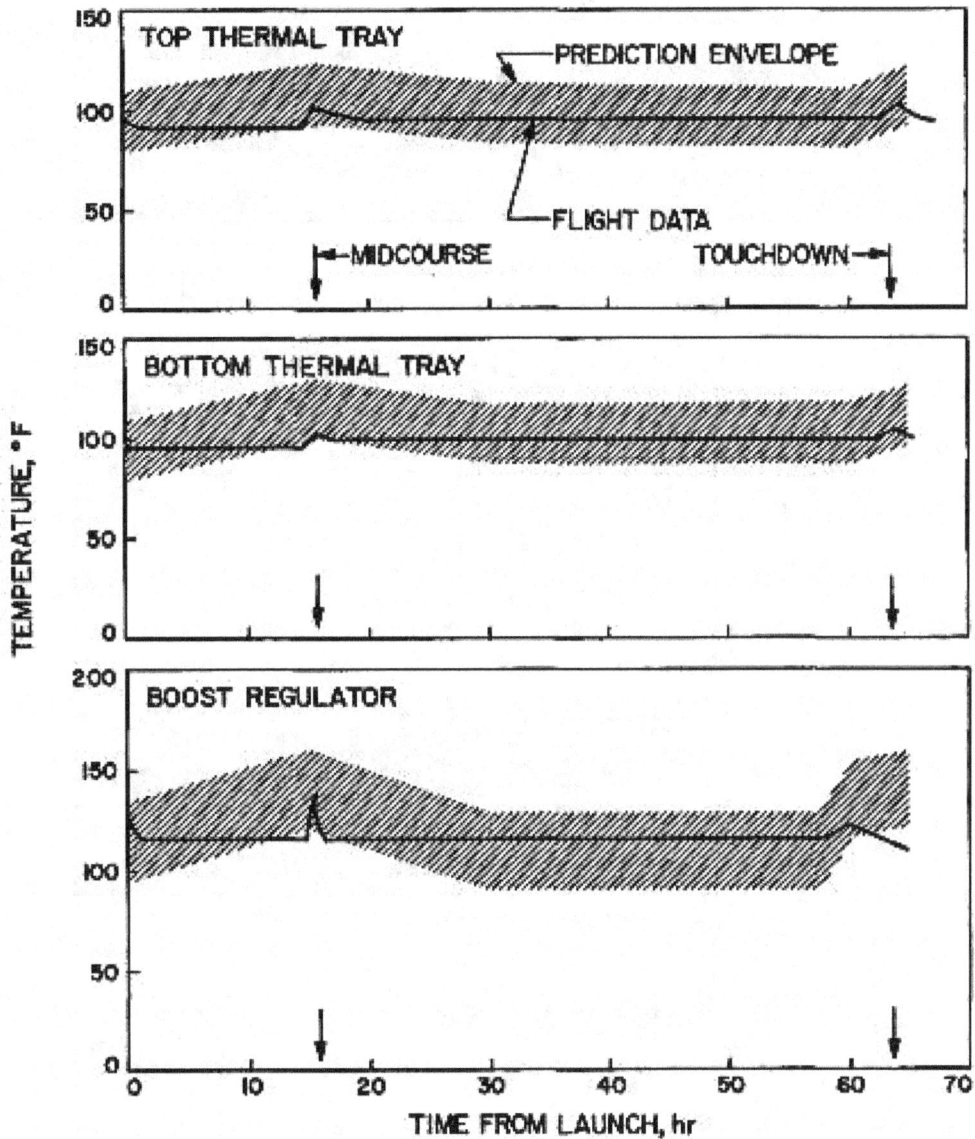

Fig. C-2. Compartment B transit temperatures

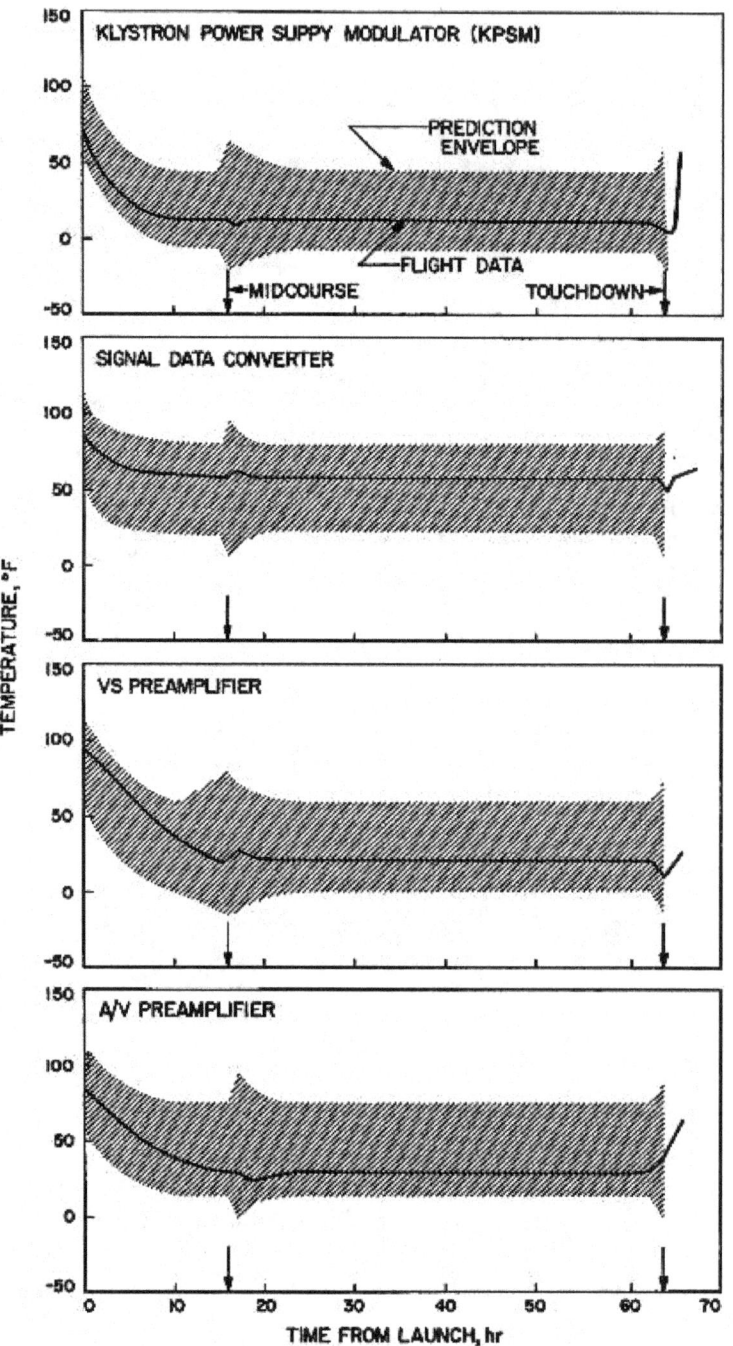

Fig. C-3. RADVS transit temperatures

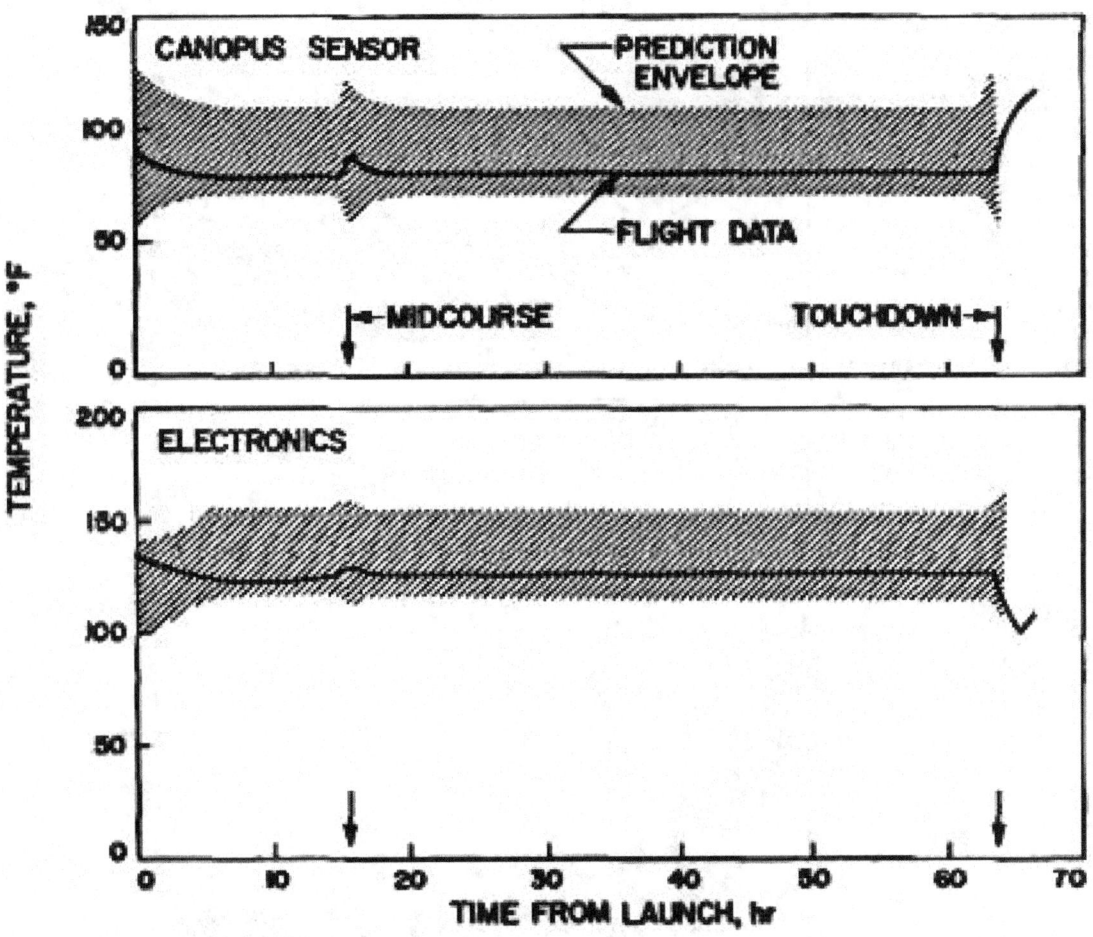

Fig. C-4. Flight control transit temperatures

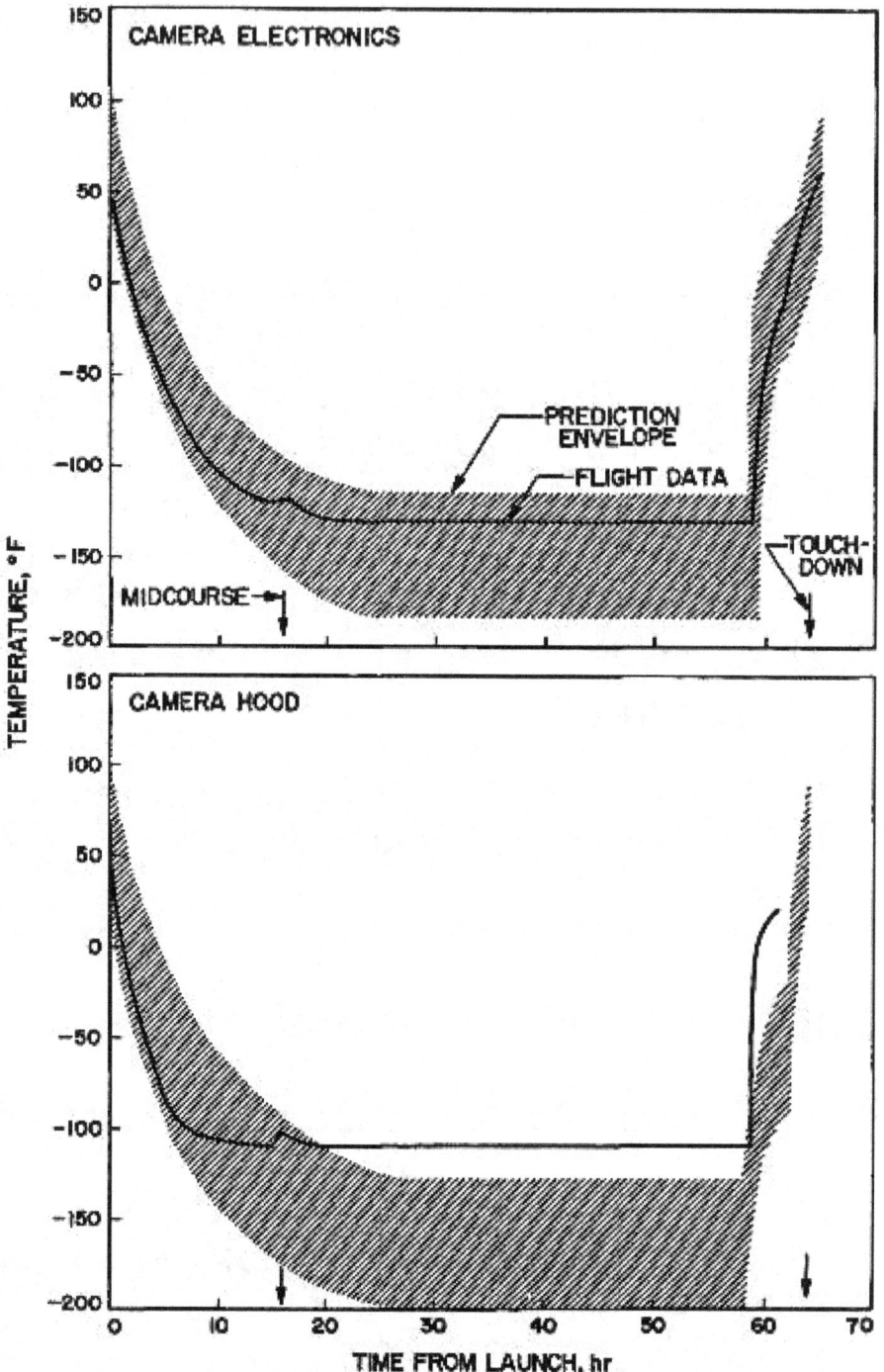

Fig. C-5. Survey TV camera transit temperatures

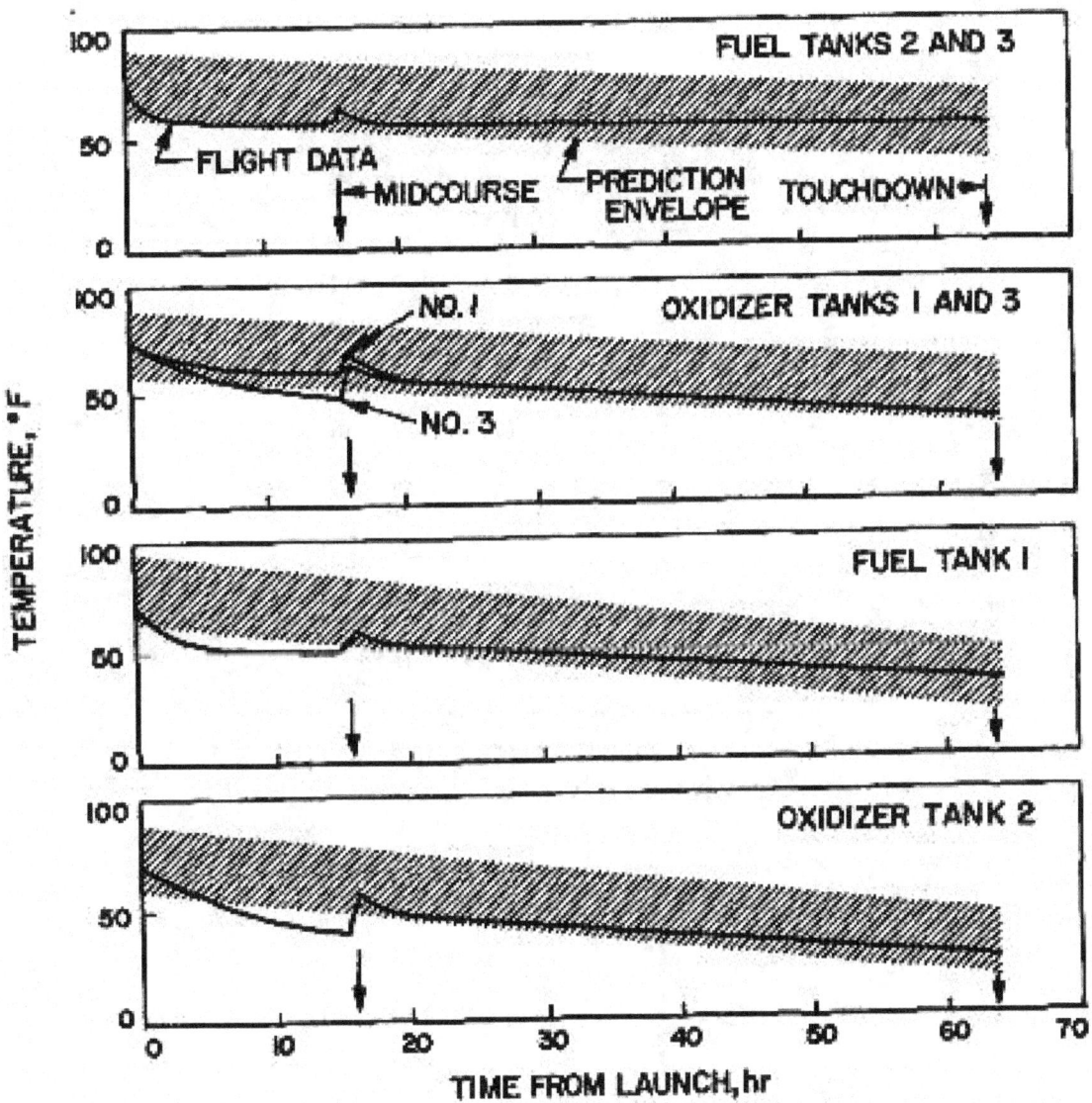

Fig. C-6. Vernier propulsion transit temperatures

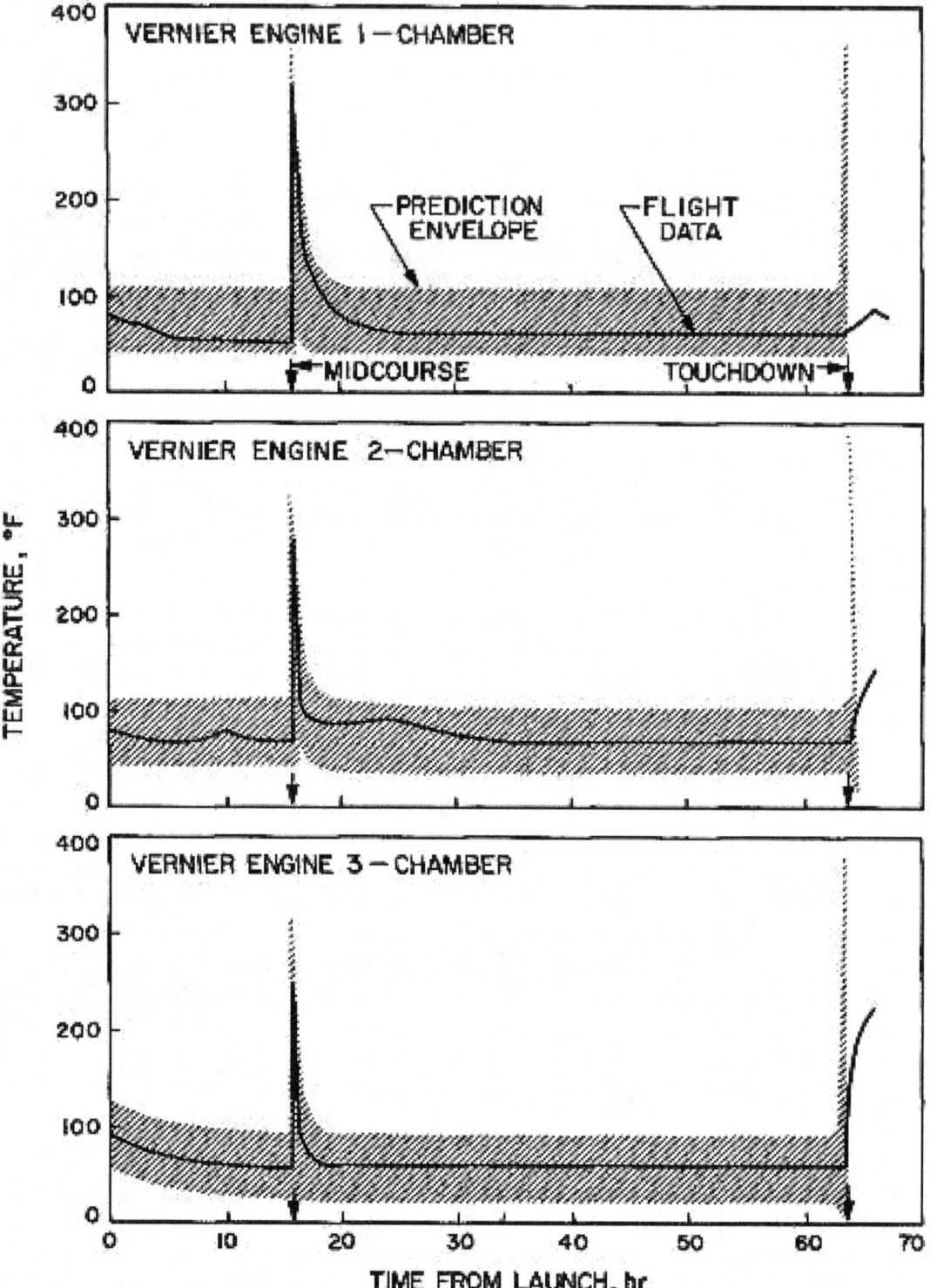

Fig. C-6 (Cont'd)

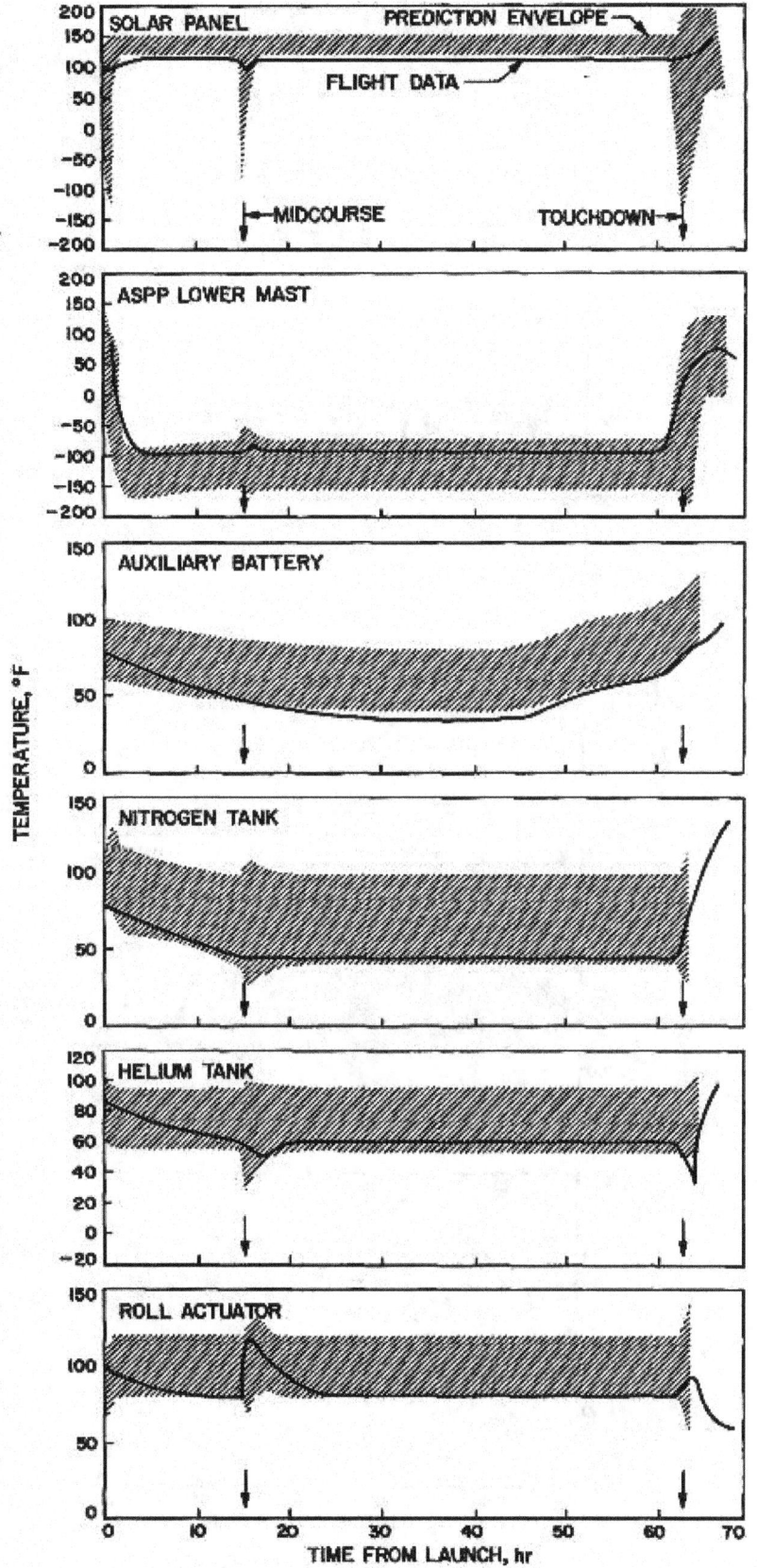

Fig. C-7. Miscellaneous transit temperatures

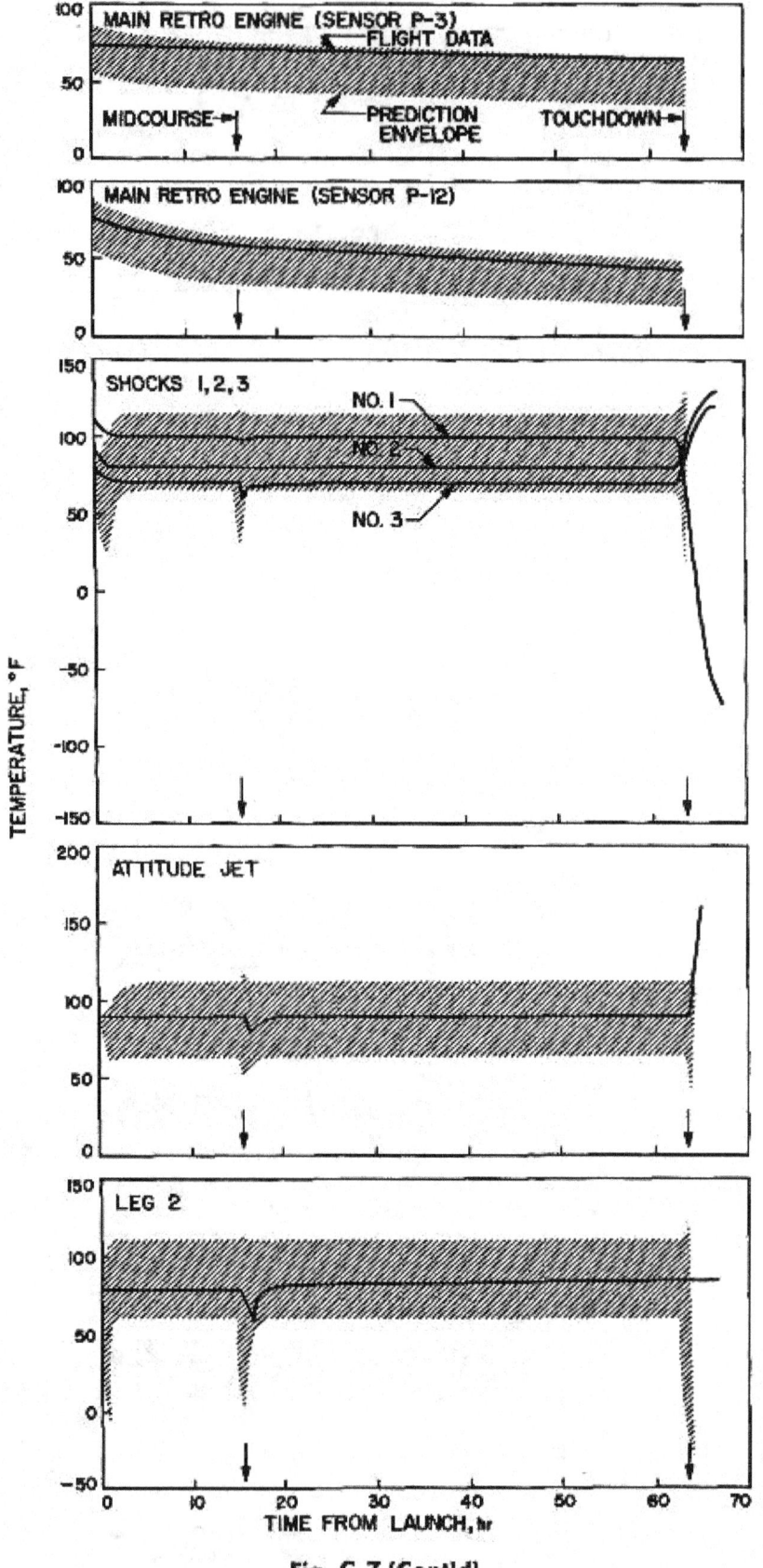

Fig. C-7 (Cont'd)

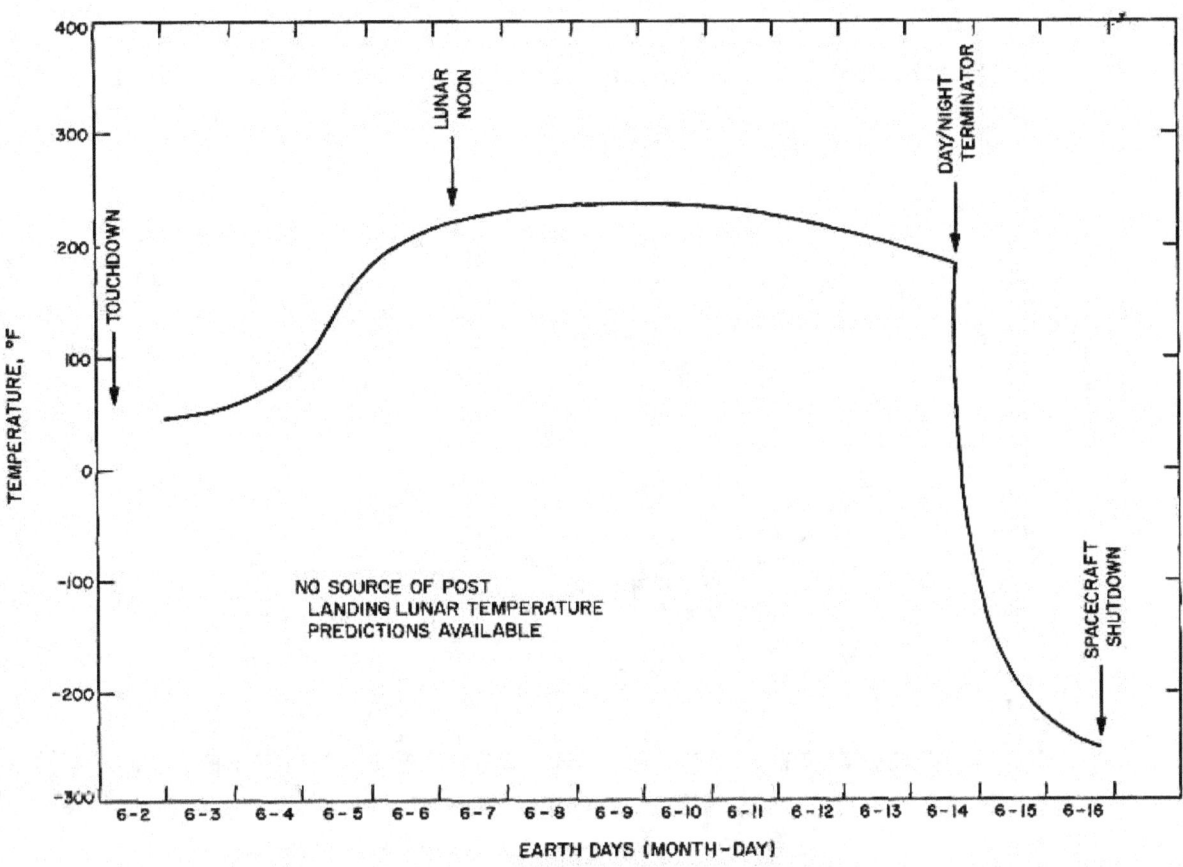

Fig. C-8. Lunar temperatures, Vernier Engine No. 1

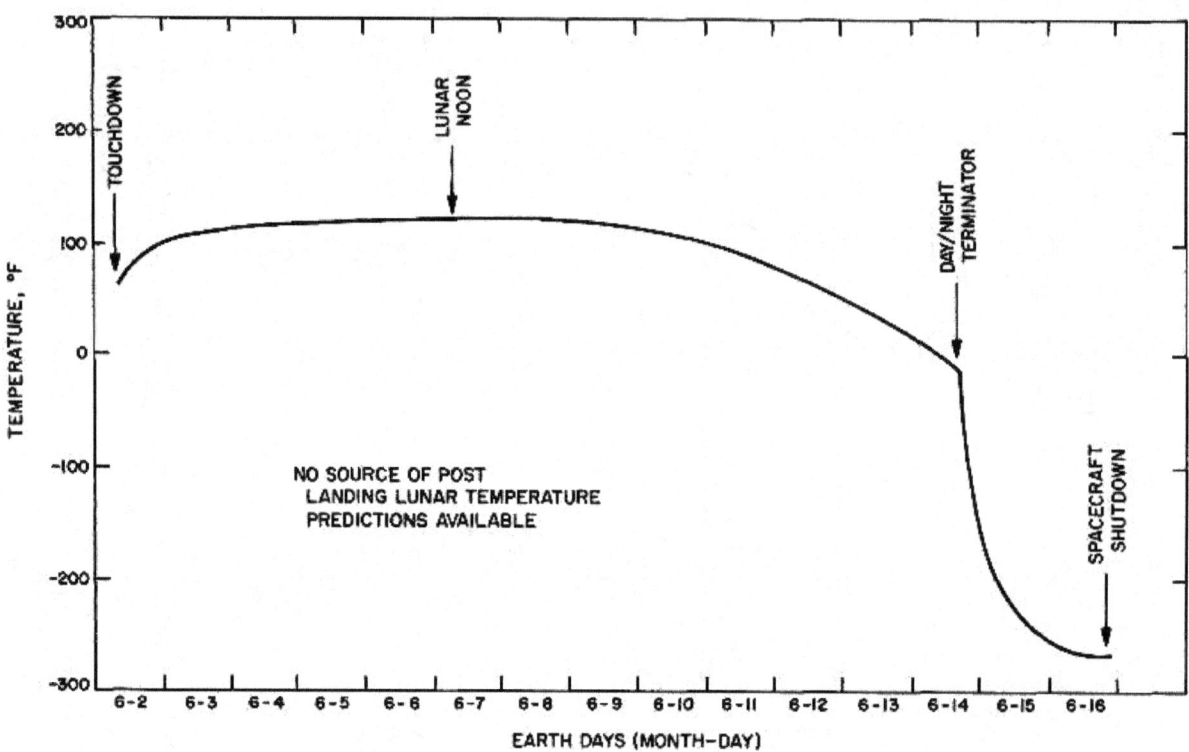

Fig. C-9. Lunar temperatures, survey camera electronics

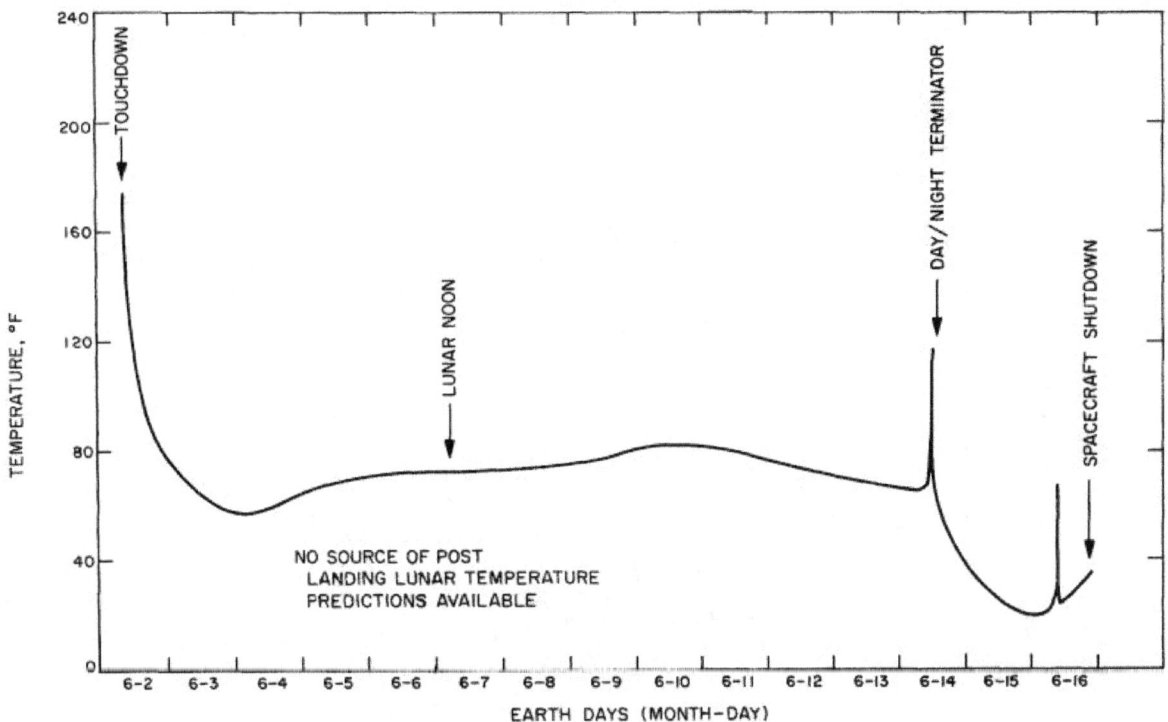

Fig. C-10. Lunar temperatures, Transmitter A

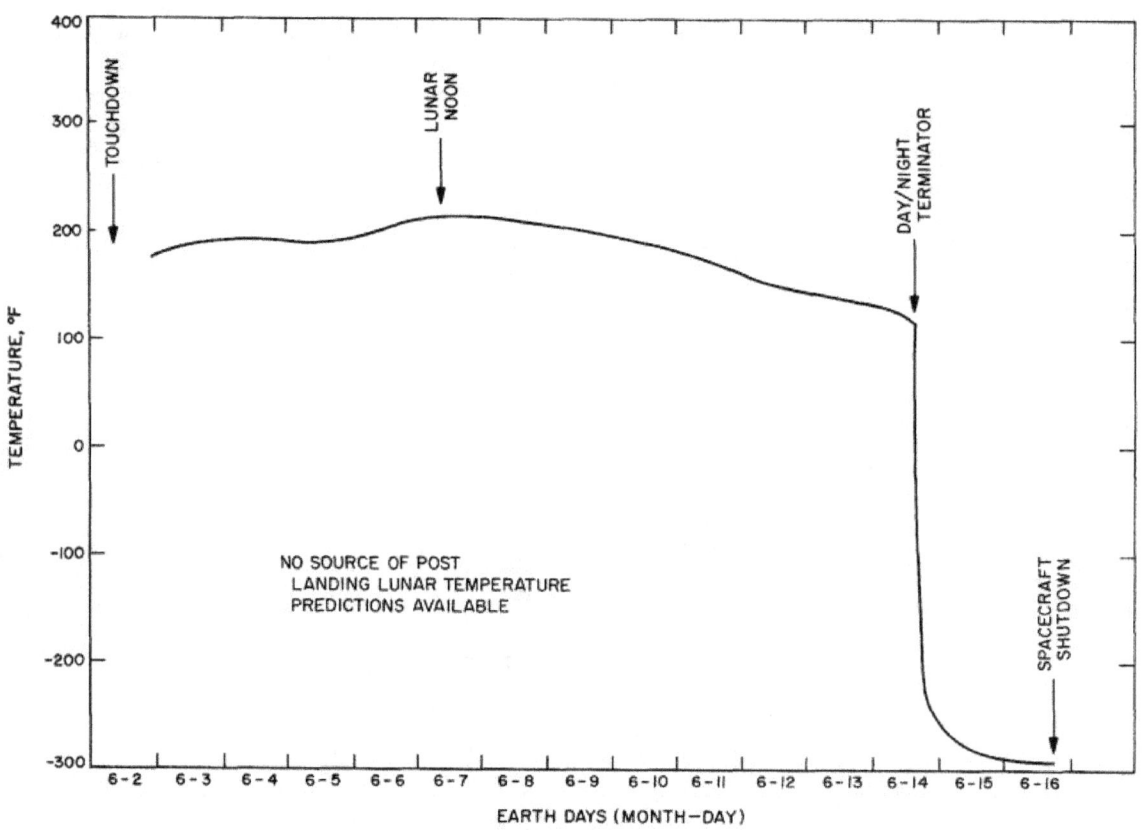

Fig. C-11. Lunar temperatures, solar panel

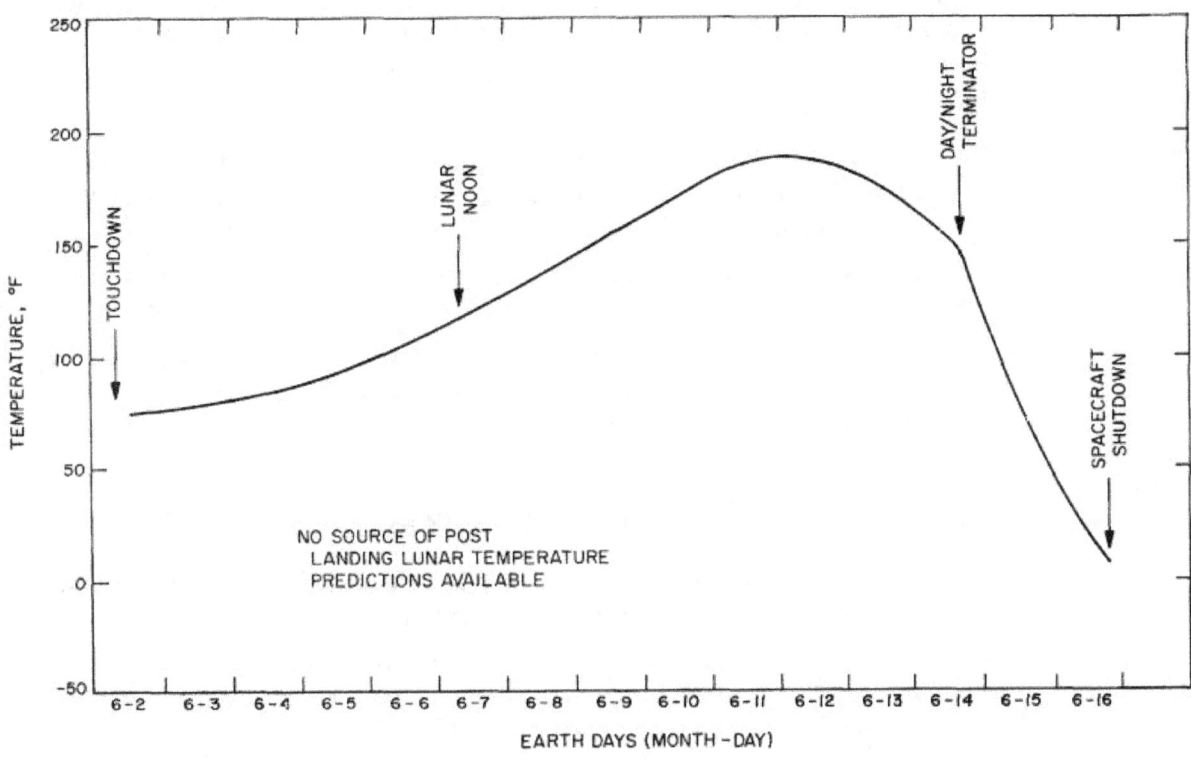

Fig. C-12. Lunar temperatures, Fuel Tank No. 1

BIBLIOGRAPHY

Project and Mission

Surveyor A-G Project Development Plan, Project Document 13, Vol. 1, Jet Propulsion Laboratory, Pasadena, January 3, 1966.

Clarke, V. C., Jr., *Surveyor Project Objectives and Flight Objectives for Missions A through D*, Project Document 34, Jet Propulsion Laboratory, Pasadena, March 15, 1965.

McGlinchey, L. F., *Surveyor Mission A Flight Test Objectives*, Project Document 32-S/MA, Rev. 1, Jet Propulsion Laboratory, Pasadena, May 19, 1966.

"*Surveyor* Mission A Postflight Analysis Meeting," minutes of meeting held at JPL July 19, 1966.

Surveyor I Preliminary Results: National Aeronautics and Space Administration Five-Day Science Report, Project Document 97, Jet Propulsion Laboratory, Pasadena, June 30, 1966.

Surveyor I–A Preliminary Report, NASA SP 126, compiled by Lunar and Planetary Programs Division, Office of Space Science Applications, NASA, June 1966.

"Space Exploration Programs and Space Sciences," *Space Programs Summary No. 37-40*, Vol. VI, for the period May 1 to June 30, 1966, Jet Propulsion Laboratory, Pasadena, June 30, 1966.

Launch Operations

Macomber, H. L., *Surveyor Block I Launch Constraints Document*, Project Document 43, Jet Propulsion Laboratory, Pasadena, June 11, 1965.

Macomber, H. L., and O'Neil, W. J., *Surveyor Launch Constraints Mission A–May/June 1966 Launch Opportunity*, Project Document 43, Addendum No. 1, Jet Propulsion Laboratory, Pasadena, May 16, 1966.

Macomber, H. L., *Surveyor A-G Mission Operations Plan (Launch Operations Phase) Mission A*, Project Document 58, Jet Propulsion Laboratory, Pasadena, May 16, 1966.

Centaur Unified Test Plan AC-10/SC-1 Launch Operations and Flight Plan Section 8.10, Report AY62-0047, Rev. B, General Dynamics/Convair, San Diego, May 2, 1966.

Test Procedure Centaur/Surveyor Launch Countdown Operations AC-10/SC-1 Launch (CTP-INT-0004L), Report AA63-0500-004-03L, General Dynamics/Convair, San Diego, May 17, 1966.

Barnum, P. W., *JPL ETR Field Station Launch Operations Plan Surveyor Mission A*, Engineering Planning Document 354, Jet Propulsion Laboratory, Pasadena, May 23, 1966.

Surveyor Mission A Centaur-10 Operations Summary, Report TR-353, *Centaur* Operations Branch, KSC/ULO, Cape Kennedy, May 24, 1966.

Surveyor A (AC-10) Flash Flight Report, Report TR-368, *Centaur* Operations Branch, KSC/ULO, Cape Kennedy, May 31, 1966.

BIBLIOGRAPHY (Cont'd)

Launch Vehicle System

Galleher, V. R., and Shaffer, J., Jr., *Surveyor Spacecraft/Launch Vehicle Interface Requirements*, Project Document 1, Rev. 2, Jet Propulsion Laboratory, Pasadena, December 14, 1965.

Atlas Space Launch Vehicle Familiarization Handbook, Report GD/C-BGJ66-002, General Dynamics/Convair, San Diego, February 15, 1966.

Centaur Technical Handbook, Convair Division, Report GD/C-BPM64-001-1, Rev. B, General Dynamics/Convair, San Diego, January 24, 1966.

Centaur Monthly Configuration, Performance and Weight Status Report, Report GDC63-0495-37 (Appendix A), General Dynamics/Convair, San Diego, June 24, 1966.

Preliminary AC-10 Atlas-Centaur Flight Evaluation, (by staff of Lewis Research Center, Cleveland, Ohio), NASA Technical Memorandum X-52212, NASA, Washington, D.C., 1966.

AC-10 Flight Review (a preliminary presentation of flight results), General Dynamics/Convair brochure, General Dynamics/Convair, San Diego.

Atlas/Centaur AC-10 Flight Evaluation Report, Report GDC-BNZ66-037, General Dynamics/Convair, San Diego, July 15, 1966.

Spacecraft System

Surveyor Spacecraft A-21 Functional Description, Document 239524 (HAC Pub. 70-93401), 3 Vols., Hughes Aircraft Co., El Segundo, Calif., November 1, 1964 (revised April 16, 1966).

Surveyor Spacecraft A-21 Model Description, Document 224847B, Hughes Aircraft Co., El Segundo, Calif., March 1, 1965 (revised June 1, 1965).

Surveyor Spacecraft Monthly Performance Report, Report SSD 68129R, Hughes Aircraft Co., El Segundo, Calif., May 21, 1966.

Surveyor Spacecraft System–Bimonthly Progress Report, mid-April through mid-June 1966, Report SS-131-SSD 68135R, Hughes Aircraft Co., El Segundo, Calif., June 23, 1966.

Surveyor Spacecraft System–Surveyor I Flight Performance Preliminary Report, Report SSD 681521, Hughes Aircraft Co., El Segundo, Calif., June 27, 1966.

Tracking and Data Acquisition

Program Requirements No. 3400, Surveyor, Air Force Eastern Test Range, Patrick Air Force Base, Fla., April 5, 1966.

Operations Directive 3400, Surveyor Launch, Air Force Eastern Test Range, Patrick Air Force Base, Fla., May 19, 1966.

Project Surveyor–Support Instrumentation Requirements Document, prepared by JPL for NASA, April 22, 1966.

BIBLIOGRAPHY (Cont'd)

Tracking and Data Acquisition (Cont'd)

Surveyor Project/Deep Space Network Interface Agreement, Engineering Planning Document 260, Rev. 2, Jet Propulsion Laboratory, Pasadena, November 22, 1965.

Pre-Flight and Post-Flight Review of the Tracking and Data-Acquisition System For the Surveyor Mission I, Engineering Planning Document 410 (preliminary), Jet Propulsion Laboratory, Pasadena, July 19, 1966.

Mission Operations System

Surveyor Mission Operations System, Technical Memorandum 33-264, Jet Propulsion Laboratory, Pasadena, April 4, 1966.

Space Flight Operations Plan–Surveyor Mission A, Engineering Planning Document 180, Rev. 5E, Jet Propulsion Laboratory, Pasadena, May 9, 1966.

Flight Path

Surveyor Spacecraft/Launch Vehicle Guidance and Trajectory Interface Schedule, Project Document 14, Rev. 2, Jet Propulsion Laboratory, Pasadena, August 13, 1965.

"Design Specification–Performance Ground Rules and Launch Periods–*Surveyor* Mission A," Specification SAO-50482-DSN-A, Jet Propulsion Laboratory, Pasadena, December 15, 1965.

"Design Specification *Surveyor/Centaur* Target Criteria *Surveyor* Mission A," Specification SAO-50439-DSN-C, Jet Propulsion Laboratory, Pasadena, January 3, 1966.

Surveyor Station View Periods and Trajectory Coordinates, SSD 5389R, Hughes Aircraft Co., El Segundo, Calif., August 1965.

Cheng, R. K., Meredith, C. M., and Conrad, D. A., "Design Considerations for *Surveyor* Guidance," IDC 2253.2/473, Hughes Aircraft Co., El Segundo, Calif., October 15, 1965.

Fisher, J. N., and Gillett, R. W., *Surveyor Direct Ascent Trajectory Characteristics*, SSD 56028R, Hughes Aircraft Co., El Segundo, Calif., December 1965.

Winkelman, C. H., *Surveyor Mission A Trajectory Design Report Revision B*, SSD 68024, Hughes Aircraft Co., El Segundo, Calif., January 12, 1966.

Pre-Injection Trajectory Characteristics Report AC-10, GDC-BTD66-032, General Dynamics/Convair, San Diego, April 1966.

Surveyor Mission A Post Injection Standard Trajectories, SSD 64085R, Hughes Aircraft Co., El Segundo, Calif., April 1966.

BIBLIOGRAPHY (Cont'd)

Flight Path (Cont'd)

Davids, L. H., and Ribarich, J. J., *Surveyor Mission A Preflight Maneuver Analysis*, SSD 68086R, Hughes Aircraft Co., El Segundo, Calif., April 1966.

Davids, L., Meredith, C., and Ribarich, J., *Midcourse and Terminal Guidance Operations Programs*, SSD 4051R, Hughes Aircraft Co., El Segundo, Calif., April 1964.

Centaur Guidance System Report, AC-10 Accuracy Analysis, BTD64-013-11, General Dynamics/Convair, San Diego, April 1966.

AC-10 Firing Tables Data, May/June 1966 Launch Opportunity, GD/C-BTD66, General Dynamics/Convair, San Diego, 1966.

Meredith, C. M., *Surveyor 1 Flight Path Analysis and Command Operations Report*, SSD 68142R, Hughes Aircraft Co., El Segundo, Calif., June 24, 1966.

Edited and Re-published by
Ross S Marshall
WeirdVideos.com